The Iwo Alamo

OTHER TACTICS/INTEL. MANUALS FROM POSTERITY PRESS:

SUPER-SQUAD: THE NOW MISSING COMPONENT
NORTH KOREA'S HIDDEN ASSETS
GUARDIAN JOE: HOW LESS FORCE HELPS THE WARRIOR
SINOLAND: THE SUBVERSION OF FREEDOM'S BASTION
AFRIQUE: A WARNING FOR AMERICA
STRATEGIC RIFLEMAN: KEY TO MORE MORAL WARFARE
GUNG HO: THE CORPS' MOST PROGRESSIVE TRADITION
GLOBAL WARRIOR: AVERTING WWIII
EXPEDITIONARY EAGLES: OUTMANEUVERING THE TALIBAN
HOMELAND SIEGE: TACTICS FOR POLICE AND MILITARY
TEQUILA JUNCTION: 4TH-GENERATION COUNTERINSURGENCY
DRAGON DAYS: TIME FOR "UNCONVENTIONAL" TACTICS
TERRORIST TRAIL: BACKTRACKING THE FOREIGN FIGHTER
MILITANT TRICKS: BATTLEFIELD RUSES OF THE ISLAMIC MILITANT
TACTICS OF THE CRESCENT MOON: MILITANT MUSLIM COMBAT METHODS
THE TIGER'S WAY: A U.S. PRIVATE'S BEST CHANCE FOR SURVIVAL
PHANTOM SOLDIER: THE ENEMY'S ANSWER TO U.S. FIREPOWER
ONE MORE BRIDGE TO CROSS: LOWERING THE COST OF WAR
THE LAST HUNDRED YARDS: THE NCO'S CONTRIBUTION TO WARFARE

The Iwo Alamo

Illustrated Edition

H. John Poole

Foreword by

Gen. Anthony C. Zinni USMC (Ret.)

Posterity Press®

Published by Posterity Press
P.O. Box 5360, Emerald Isle, NC 28594
(www.posteritypress.org)

Cataloging-in-Publication Data
Poole, H. John, 1943-

The Iwo Alamo
Includes bibliography and index.
1. Infantry drill and tactics.
2. Military art and science.
3. Military history.
I. Title. ISBN: 978-0-9638695-0-0 2020 355'.42
Library of Congress Control Number: 2020915747

Edited by Dr. Mary Beth Poole

First printing, United States of America, October 2020

For all U.S. infantrymen interested in knowing what it would now take — at the squad level — to seize a modern-day place like Iwo Jima.

Cover art © by Col. Charles A. Waterhouse USMCR
an enlisted veteran of the WWII Battle for Iwo Jima

Born on 22 September 1924, Charles Waterhouse had been a member of the class of 1942 at Perth Amboy High School in New Jersey. In August 1943, he enlisted in the United States Marine Corps to eventually join Company C, 5th Engineer Battalion, 5th Marine Division, Fleet Marine Force.[1] Upon acquiring the rank of PFC, he landed with the first assault wave at Iwo Jima on 19 February 1945. When the beach finally came under heavy and continuous enemy fire, Waterhouse was wounded, suffering permanent nerve damage to his left hand. For that battle-inflicted injury, he would later receive the Purple Heart Medal.[2] Engineers played a key role in the initial assault on Iwo Jima. Right behind the beaches were machinegun-covered minefields and pillboxes. It was the job of engineers to clear a path through those minefields and (in demolition teams attached to the infantry) destroy heavy pillboxes.

CONTENTS

ILLUSTRATIONS

Maps:

Figures:

Foreword

Every Marine knows about the battle for Iwo Jima. It is the iconic battle of World War II immortalized in the photo of the Marines raising the American flag on Mount Suribachi. I became more intimately familiar with the battle in 1987-89 while a regimental commander on Okinawa. Our chief of staff directed the colonels in the division to each take on an assigned battle from the World War II Pacific theater, research the battle assigned in detail, walk the battlefield, and then conduct groups of Marines on staff rides over the battlefields to discuss lessons to be drawn from the fighting. Consequently, I made a number of trips to the island, even some with Japanese military officers and members of the Japanese Iwo Jima Association, one of whom had been stationed on there and left just before the battle commenced.

Since that time, I have continued to research the battle and interviewed a number of Marines who fought in the battle. Their stories are remarkable tales of bravery, camaraderie, survival, and ingenuity unfolding in the hell of ferocious close combat. It is very easy to acknowledge all this but, as many military historians and analysts often do, not see the value in capturing important lessons from the battle that are applicable today.

The wisdom of the strategy that led Admiral Nimitz to Iwo Jima is often argued and the conduct of the operational level is easily dismissed as just a brutal frontal clash of forces not worthy of analysis. There is, however, a richness in the lessons that can be drawn from an individual and small unit study of the battle and what that can inform us as to the skills, tactics, training, leadership, and warrior ethos necessary for surviving and winning in an intense, close combat environment.

The perfect person to capture those lessons is John Poole. His masterful series on war fighting has demonstrated his fundamental knowledge of what it takes to prevail in today's potential conflicts. He has an uncanny ability to draw out the lessons of history and filter them through his own impressive set of experiences and lay out how we need to develop warriors to meet the current threats to our security.

Too often we get caught up with grand strategy, national security policy, and the allure of advanced technology and need to be reminded that we still face the challenge of close combat and the need for junior officers and NCOs to carry the day. We saw this in the heavy reliance on so called "shock and awe" in the invasions of Iraq and Afghanistan that led us to underestimate the requirement for infantry units to achieve victory. The momentum was lost and the scramble to add infantry forces in a "surge" came too late for an overwhelming victory. This mistake cost us lives and mission failures from the outset.

In this book, John Poole gives us a valuable appreciation and guide as to what it takes to create the battle skills and character to meet the challenges of the close fight that still remains as the critical part of winning wars.

GEN. ANTHONY C. ZINNI USMC (RET.)
FORMER HEAD OF CENTCOM

PREFACE

February 2020 marks the 75th anniversary of one of the toughest battles in U.S. history—that for the tiny Pacific island of Iwo Jima. This 36-day struggle has so far inspired hundreds of books and articles. Why would somebody two years old at the time feel obligated to revisit its details? The reason is simple. America's military has yet to assimilate the primary lessons of that epic ordeal. Such an oversight not only disrespects those who died there, but puts all modern-day GIs (government issues) at risk

In *The Iwo Alamo,* this researcher has tried to recapture what U.S. infantrymen experienced late in WWII for three reasons: (1) to further honor them; (2) to see if recent insights into enemy tactics have enhanced their accomplishment; and (3) to show how little today's technology would have helped them. The human factor on Iwo was to play a much larger role in its capture than any equipment. The resolve and adaptability of frontline troops would turn the tide of what could have easily become a major disaster. But, in the process, many fine young Americans would have to face their own personal Alamo, (See Figure P.1.)

Because of the massive size of the invasion force, the actions of U.S. battalions and companies have been recorded—but not of their individual squads. Armchair historians have only been able to dig up a few tiny-element exploits. So, *The Iwo Alamo* uses bravery citations to expand upon the overviews of existing government documents. In this way, it has produced heretofore missing detail into the frontline realities of the battle.

In today's fast-paced world, most U.S. citizens and service personnel have come to believe that all military skills become more refined over time. Yet, one of America's greatest wartime leader's Gen. George S. Patton once warned, "To be a successful soldier, you must know history."[1] But this is no easy task, because official battle chronicles seldom discuss foe accomplishments, and post-war bureaucracies regularly disregard lessons learned. Some small-unit

Figure P.1: Death Was Never Far Away on Iwo
(Source: "Flotsam and Jetsam ," by Charles Waterhouse, from "Closing In," by Alexander, Marines in WWII Commemorative Series, HQMC, 1994, p. 18.)

stratagems may improve with the latest paraphernalia, but most atrophy (as with a crutch). Most military trainers realize that Rogers' Rangers (of the pre-Revolutionary-War era) could have more stealthily approached a prepared enemy position than contemporary U.S. commandos. Additional gadgetry may help today's "special operator" to shoot exposed defenders, but it does nothing to obscure his presence.

Failing to refine standard operating procedures can have serious consequences in combat. To determine whether the U.S. Military could still win an all-out war, one must try to imagine how current GIs would fare at the Iwo Jima of 1945. All current conditions must be considered: (1) female commanders and fellow fireteam members; (2) cell-phone-oriented riflemen with limited woods skills; (3) less cohesion through two-man rooms; (4) officer-centric units no longer reliant on Staff Noncommissioned Officer (SNCO) help; (5) less emphasis on physical conditioning; and (6) most close-combat training through virtual reality.

All other Posterity Press titles have focused on the mechanics of small-unit maneuver. This one will place the reader at the very heart of the action. The experience should prove as exciting as it is insightful. But, war is not one of man's more noble inventions, so there will be no feeling of contentment. God has already provided for all the GIs who did their duty on Iwo. It's now up to America's modern-day service personnel to try to do as well.

LT.COL. H. JOHN POOLE USMC (RET.)
CAREER INFANTRYMAN AND
FORMER FLEET MARINE FORCE GY.SGT.

Acknowledgments

This work, like the 19 others from Posterity Press, has been for God's glory. When Jesus said, "I am the truth, the way, and the light," he may have been speaking literally as well as figuratively. In other words, all that is not completely consistent with His teachings may be misinformation. The Bible goes so far as to imply that everything other than a simple "yes" or "no" to a moral question contains nonproductive exaggeration.

> Let your yes mean yes, and your no mean no. Anything more than this comes from the evil one.[1]
>
> — Matthew 5:37, *Common English Bible*

Top-down governments will often "spin the truth." They don't do so to deceive anyone, just to put what happened into the most favorable light. But, where the lives of one's soldiers are concerned, that's no blessing. Too little frontline detail can lead to small-unit inadequacies in some future conflict.

To revisit the Battle of Iwo Jima, one must focus more on the spiritual than the horrific. As in all wars, the enemy on that tiny World War II (WWII) island had been an active participant. Yet, to this day, the American way of war is still to so overwhelm a foe with firepower that his tactics make no difference. With any Asian adversary, that can prove disastrous for two reasons: (1) the Sun Tzu tradition of deception; and (2) where state-of-the-art small-unit maneuvers come from. For most squad scenarios, the most productive stratagems had been collectively developed by all big Eastern armies before the end of WWII.[2] That's why the Marines had such a hard time taking Iwo. After underestimating the opposition's cunning, their leaders should now be applauded for allowing enough frontline initiative to overcome it. This joint effort turned out to be a lasting testament to the Holy Spirit within. That's why uncommon valor became a common virtue on Iwo.

Part One

Accomplishing the Impossible

"And when he gets to Heaven, Saint Peter he will tell:
'Another Marine reporting, sir—I've served my time in Hell!' "

(Source: Possibly a revised version of "Our Hitch in Hell," as found in the book "Mud and Stars - An Anthology of World War Songs and Poetry," by Dorothea York © 1931.)

1 The Assigned Task

- Had U.S. planners foreseen all obstacles to invasion?
- What would the landing force face the first day on Iwo?

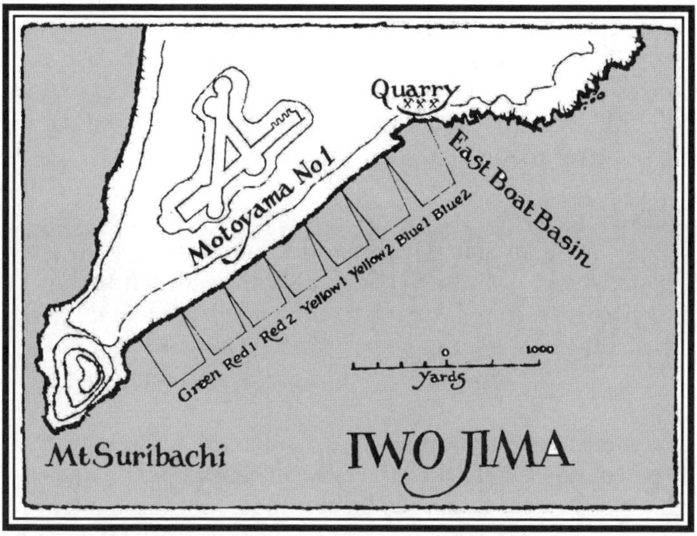

(Source: Bottom part of "Map of Iwo Jima," by Col. Charles A. Waterhouse, "Fortitudine," vol. 24, no. 4, Spring 1993, Hist. Div., HQMC, possible copyright ©.)

The Context of a Historical Event Can Expand with Time

What happened at Iwo Jima has become much clearer to U.S. tacticians as they came to better understand the Oriental thought process and deceptive style of war. Most Americans have little personal experience with a sub-surface defense. So, fully to understand what their WWII Leathernecks had overcome on that tiny Pacific isle, they may need an Asian opinion.

> [U.S.] planners drew up the details of Operation Detachment, estimating the island would be cleared of defenders in a week.

> U.S. intelligence, however, had badly underestimated the tactics of Lieutenant General Kuribayashi, who had eight months to build fortifications that included many kilometres of underground tunnels, bunkers and strongpoints.[1]
>
> — *South China Morning Post,* 2 June 2019

That eight-square-mile Japanese fortress had been intermittently bombed and shelled since June 1944. Starting on 8 December, it was additionally plastered with 6,400 tons of bombs and 22,000 shells from U.S. planes. And then, right before the landing, American ships bombarded it—for another three days—with a nearly continuous stream of large-caliber munitions.[2] But, the beach fortifications received most of this final attention. Only slightly affected were the carefully hidden emplacements of Kuribayashi's protective arrays atop the central plateau. (See Map 1.1.)

> On D minus 3, when naval preliminary bombardment began, 750 . . . [major defensive installations] could be counted [from the air]. All the firepower of the mighty fleet had to be concentrated on destroying beach defenses alone. . . . [So, inland] Iwo would have to be reduced the hard way.[3]
>
> —*Third Marine Division,* by Aurthur and Cohlmia

Reconnaissance swimmers and Underwater Demolition Team (UDT) personnel had checked the eastern beaches for obstacles. But, only the area behind them had been mined, and those naval scouts had been looking for man-made barriers from just offshore to the waterline. Though undoubtedly aware of the Asians' predisposition toward deception, U.S. Intelligence had concluded in December 1944 that there might be an initial flame-based deterent.[4]

> Partially buried gasoline drums were observed [by a U.S. submarine] at the water's edge on both the east and west beaches. The irregular pattern led intelligence experts to believe that they might be some kind of beach obstacles. This belief was strengthened by [the] capture of Japanese documents giving detailed instructions for employment of burning gasoline to check hostile landing forces.[5]
>
> — *Iwo Jima: Amphibious Epic*
> Historical Branch, HQMC

That's when overemphasis on this possibility had led to a failure to notice natural beach obstacles. Through unusual coastal dynamics, ostensibly firm sand at Iwo Jima had actually been a thick emulsion. And just behind that overly soft shoreline had lain a steep terrace of loose volcanic ash. The combination would prevent any new occupant, mechanical or otherwise, from acquiring either a firm footing or protective hole. And to the heavily-loaded GIs who would have to win this fight, it was psychologically debilitating.

> We had to crawl up the deep sand. They had no idea what was on that island.[6]
>
> — Iwo Jima veteran with tears in his eyes
> during a 2018 Interview

> Marines of the landing force, many of them weighted down with more than 100 pounds of weapons and other gear, found themselves floundering in a sea of soft volcanic ash that all but precluded their ascending that 15-foot seawall. Almost immediately, the Marines sank up to their ankles into the loose ash that tugged at their feet and made all forward movement a strenuous undertaking.[7]
>
> — *History of Marine Corps Operations in WWII*
> Historical Branch,Hdqts.Marine Corps (HQMC)

All Japanese initiatives so far may have been actually part of an evolutionary defense. Kuribayaski had known about the tactical innovations of the WWI Germans and Mao. So, as a cultural descendent of Sun Tzu, he had laid a trap for a foe who—to this day—over-relies on firepower and sees the world through ordnance-colored glasses.[8] But in February of 1945, the Americans had Right on their side, and the clever opponent didn't. From the cliffs above the Quarry, four big coastal-defense guns had—with Kuribayashi's permission—opened fire on landing-craft gunboats protecting the U.S. reconnaissance party. Otherwise, that brace of oversized Japanese cannons would have still been there on D-Day to take the entire U.S. debarkation zone under direct fire from a flank.[9]

The Lull Before the Storm

Then came the largely unopposed landing. For almost an hour,

Figure 1.1: First Waves Took Only Limited Fire Coming Ashore
(Source: U.S. Army Center of Military History, posters, image designator 1-25-49.jpg, "Beach Assault," by Amos Sewell, public domain.)

the defenders of Iwo held most of their fire against the newly arrived intruders.[10] (See Figure 1.1.) Because of the island's unique shape, Japanese supporting-arms observers enjoyed a near-perfect view of not only its beaches, but of the entire isthmus.[11] The U.S. landing plan had called for enough assault waves to be ashore within 45 minutes to total 9000 Marines.[12] After at least 8000 members of two divisions had made it in and experienced difficulty leaving the beach area, Nipponese gunners opened up with all the long-range

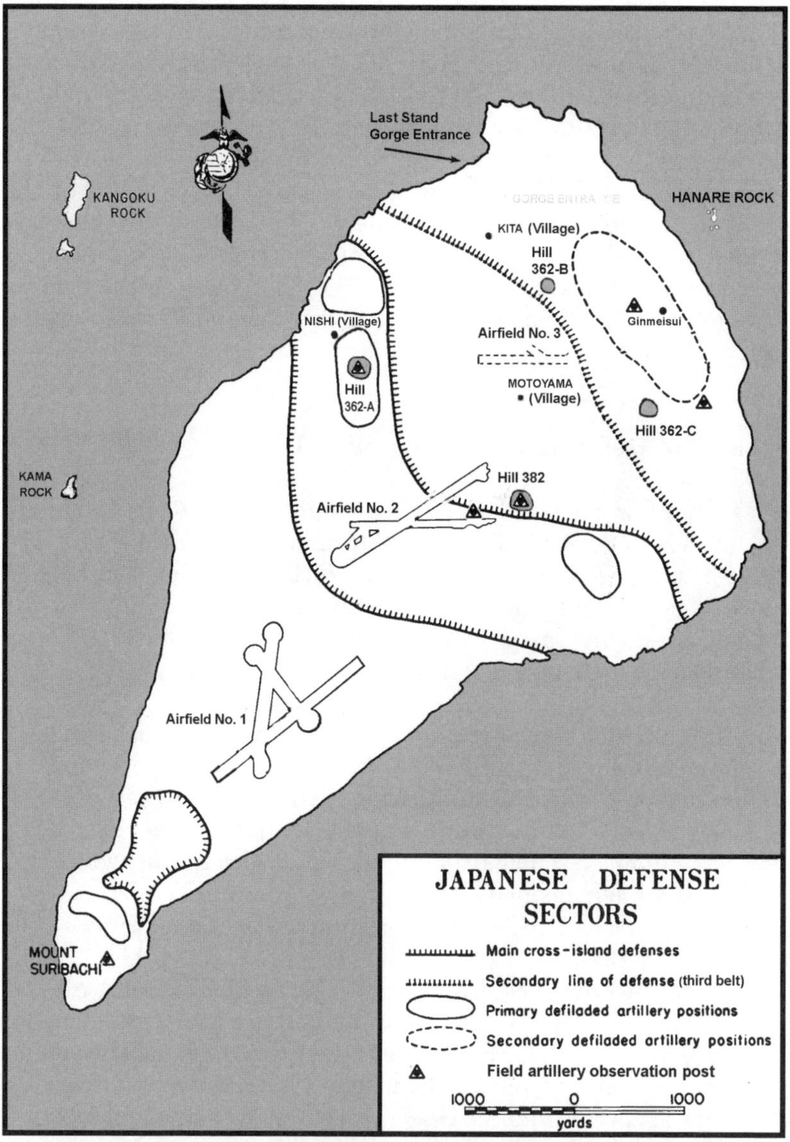

Map 1.1: Initial U.S. Estimate of Defense Line Locations
(Source: "Closing In: Marines in the Seizure of Iwo Jima," by Alexander, Hist. & Museums Div., HQMC, 1994, pp. 5, 8.)

firepower they could muster.[13] They were trying, of course, to get maximum effect from a limited supply of ammunition. Their primary target—that Yank-crowded landing area—was now some 3000 yards long (Blue 2 was not being used) and nowhere wider than 300 yards (due to the Navy's preplanned rolling barrage).[14] Defensive planners had covered this quarter-square-mile kill zone with numerous artillery and mortar preregistrations that could be easily adjusted onto a lucrative target. Half of that area could also be reached by anti-aircraft guns on Suribachi. and the other half from similar weapons atop the Quarry cliffs. (Look again at the cover map.) Around the same time as this indirect-fire barrage, hidden machineguns to the left, front, and right of the advancing Marines completed the ambush.[15] (They had been on the leading edge of either Airfield No 1 or the Suribachi and first Motoyama defense zones on Map 1.1.)

What Had Delayed That Move Inland

Only a few of the 69 troop-laden amphibious tractors (amtracs) in each assault wave had managed to climb the terrace of ash, and most of their accompanying mortar/rocket vehicles had bogged down at the soft shoreline.[16] The ocean was rough that day. So, the incoming assault waves soon created their own obstacles. That's when Kuribayashi struck.

> The surf zone was narrow but violent. The soft, black sand immobilized almost all the armored . . . vehicles . . . and bellied up some of the amtracs. . . . Once the beaches were choked with landing craft and the steep terraces clogged with infantry, Kuribayashi fired signal flares, after which the defenders opened up with heavy ordnance.[17]
>
> — *The National Interest,* 15 December 2018

> At the water's edge amtracs, LCMs and LCVPs were hit, burned, broached, capsized, and otherwise mangled. The loose, black volcanic cinders, slid past the churning tires of wheeled vehicles, miring them axle-deep; the steep terraces blocked egress from the beach and extensive minefield took a heavy toll. Debris piled up everywhere.

> Wounded men were arriving on the beach by the dozen, where they were not much better off than they had been at the front. There was no cover to protect them and supplies of plasma and dressings ran low. The first two boats bringing in badly needed litters were blown out of the water. Casualties were being hit a second time as they lay helpless, under blankets, awaiting evacuation to ships (Connor, *The Fifth Marine Division,* 53).[18]
>
> – *History of Marine Corps Operations in WWII*
> Historical Branch, HQMC

"Despite the enemy fire, congestion at the water's edge, and initial confusion accompanying the landings, men and supplies continued to pour ashore." Largely due to the initial deluge of enemy fire into a such a confined area, the invasion force was to suffer some 2400 casualties before the end of the first day.[19]

An Underground Citadel Existed

Iwo Jima had been continually surveilled by U.S. aircraft before the landing. During the pre-invasion bombardment, the Japanese had only visibly emerged from their bomb shelters to repair the runways of Airfield No. 1.[20] Concrete blockhouses could be clearly seen just above the eastern beaches. But, those had been built before Gen. Kuribayashi arrived,[21] and would subsequently serve as a diversion for Allied bombers. His plan was to defend farther inland through a succession of "soft" strongpoint belts. The exact trace of those belts had been carefully obscured, and was thus not apparent from the air. (Look back at Map 1.1.)

> [T]he eve of the scheduled landing . . . the latest photographs . . . seemed to indicate that the principal defense installations on and behind the eastern beaches . . . had been destroyed or heavily damaged. To effect this destruction . . . inland defenses had been neglected and many were untouched.[22]
>
> –*WWII History of 5th Marine Division,* by Connor

The full extent of that inland array has only recently been realized.

> [Inland,] more than 1000 gun [covered artillery] emplacements and [mostly hidden] pillboxes had been connected by miles of tunnels with underground headquarters and hospitals.[23]
>
> — *Marines TV,* 14 March 2012

> After constructing 11 miles of fortified tunnels that connected 1,500 rooms, artillery emplacements, bunkers, ammunition dumps, and pillboxes, the 21,000 Japanese defenders could fight almost entirely from underground.[24]
>
> — *The National Interest,* 15 December 2018

What All Those Connector Conduits Made Possible

Most of those inland strongpoints had immediate access to extensive below-ground compartments. Besides tiny exits and ventilation shafts in the surrounding rocks, those sub-surface networks had enough tiers and tunnels to guarantee breathable air under almost every circumstance. This made them impossible to neutralize with surface initiatives.

> Some underground positions now boasted five levels.[25]
>
> — *The National Interest,* 15 December 2018

The sub-surface passageways also made those strongpoints easy to abandon at the last moment for a fallback position. In this way, Kuribayashi could get maximum use out of a limited manpower and logistics pool. For a society of "big-picture surface dwellers who never retreat," the impregnable nature of such a fluid, below-ground arrangement was difficult to comprehend. No degree of bombardment could much affect it. And even late in the battle, Marine units could not take the time to investigate every ground opening they encountered, though most could be reoccupied from below.

> [A]erial photographs didn't show all the [tiny] positions that were inside the rocks. They were still there when we landed.[26]
>
> — Commanding Officer of E/2/25

Overall Impact of the Preparatory Bombing and Shelling

While the pre-invasion bombardment had made the beaches easier to occupy, it had largely ignored the inland emplacements.[27] That may have increased the extent to which the strongpoint belts would be refined—through more bomb shelters and connector tunnels.

> One of the results of American air attacks and naval bombardment in the early summer of 1944 had been to drive the Japanese so deep underground that eventually their defenses became virtually immune to air or naval bombardment.[28]
>
> – *History of Marine Corps Operations in WWII*
> Historical Branch, HQMC

The shattering impact of all that heavy U.S. ordnance had also helped to clear fields of fire for Japanese machinegunners. To properly protect their respective strongpoints. they needed interlocking bands of grazing fire. That took more level and foliage-free ground.

> [T]he powerful invasion barrage . . . gave a levelling *[sic]* effect to . . . the jumbled ground formation which later formed a perfect natural defense area.[29]
>
> —*WWII History of 5th Marine Division,* by Connor

Full Scope of the Japanese Fire Plan

At Iwo (as at Peleliu), the Japanese would try a different way of resisting invasion. They would make no attempt to stop the initial landing, or to conduct a massive *banzai* attack against its participants. Instead, intruder units would be so heavily "attrited" during any move inland that they would experience a significant loss of combat power. (Or as Santa Anna's troops at the hands of Davy Crockett and Jim Bowie, they would become so bogged down as to be incapable of further advances.)

A large part of the Nipponese strategy was to blanket every square yard near a strongpoint belt with some form of lethal fire.

That way, the Americans could not even approach one without unacceptable casualties. In addition, all defensive stands would be like an ambush.

> [During the move north], we never saw a jap but fire was coming out of every hole and every rock.[30]
>
> — U.S. Dept. of the Navy Documentary

> We hardly ever saw an enemy. The Japanese had every inch of ground covered by fire.[31]
>
> — Navaho Marine veteran
> "Japanese Codetalkers," *History Channel*

Effect of the U.S. Blockade

The U.S. Navy had blockaded Iwo Jima for several months before the landing, but a number of small Japanese boats still gotten through.[32] Besides food, some may have been carrying water.

> [B]y February [1945], troops on Iwo had seventy days' supply of rice, plus other rations which could be made to last for two [more] months.[33]
>
> —*WWII History of 5th Marine Division,* by Connor

The Potable Water Situation on Iwo

This U.S. attack objective has been generally thought to be without a natural water supply. But, in the summer of 1944, there had been an oak forest near Motoyama Village. After the invasion, foragers had tried to reach a well south of Hill 362-A. Marine engineers had come up with "hot mineral water" while digging a well near the landing beach. And both the Seabees and later occupiers would drill usable wells on the isthmus.[34] So, there was ground water on the island.

Iwo Jima translates as "Sulphur Island," so well water may not have tasted or smelled all that good. That's why local residents had preferred what fell from the heavens. Just as the desert rajas had done with their mountaintop fortresses during the Mughal invasion of India,[35] they had channelled rainwater runoff into cisterns. Around Motoyama Village, there had been hundreds of interconnected surface cavities carved out of the soft stone.[36] Then, for several months before the invasion, Japanese engineers expanded that system, to possibly include large sub-surface reservoirs. Working parties are known to have dug wells and built concrete storage facilities. A distillation plant was constructed near the East Boat Basin with a 10,000-gallon tank. But the Japanese knew those distinctively white structures would attract bombing and shelling. Elsewhere hundreds of condensers were less obviously built to extract water from steam vents in the two volcanoes. While none of the wells near Motoyama Village would come to fruition, some of the others may have. And U.S. forces were about to make a mistake in judgment. According to a recent Quantico study, they would not fully target the original collection system. In all likelihood, they had mostly gone after decoy structures.[37]

> The large [surface] concrete cisterns were either wholly or partially destroyed by our gunfire early in the assault phase.[38]
>
> — *Intel.Report,*G-2,TF (Task Force) 56, April 1945

In other words, at the time of the invasion, the defenders of Iwo could have had enough water on hand at key locations for the entire 20,000-man force to hold out for several months. As the battle progressed, they simply ran into distribution problems. Nor were there as many troops to support. Yet, no other historical analysis has pursued this line of thought.

Today, the rainy season on Iwo lasts from April to September. So, Kuribayashi's engineers could have collected a rather large and durable supply of water after the bombing started in June of 1944.[39] Probable spots for sub-surface reservoirs would have been at the base of Suribachi (fed by Airfield No. 1 runoff), center of the 2nd Defense Line (fed by Airfield No. 2 runoff), and near the various fallback positions (fed by plateau runoff).

> The average . . . precipitation for the year in Iwo Jima is 54.4" [inches] . . . . The month with the most precipitation . . . is July with 7.1" . . . , with the most rain occurring in August with 9.6 days of rain.[40]
>
> — *Weatherbase.com* for Iwo Jima, Japan, 2020

Just prior to D-Day, "heavy rain" had hit Iwo.[41] Then, it rained for three days after the landing, and again at the beginning of March.[42] There was, in addition, a suspicious entry on a post-war Japanese map.[43] Near the northeast coast had been something called "Ginmeisui (Ginmyosui)." (Look back at Map 1.1.) The only known translation of the term from that period is "a fountain on Mount Fuji."[44] That fountain may have been a natural spring, or the veiled reference to a homeland transit point.

Extent of Damage to the Nipponese Command and Control

By February 1945, most of the enemy strongholds on Iwo Jima had been linked by subterranean or entrenched wire, with a radio backup. So, the final pre-landing bombardment would have little effect on Kuribayashi's ability to coordinate his defensive efforts. Nor would it impede the adjustment of preregistered Japanese fires. (See Figure 1.2.)

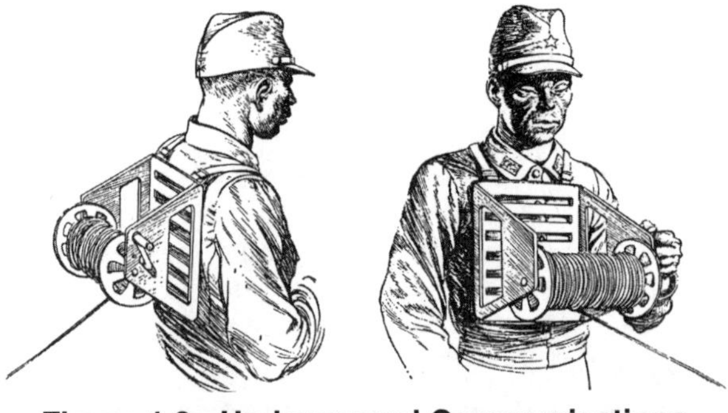

Figure 1.2: Underground Communications
(Source: "Handbook on Japanese Military Forces," TM-E 30-480 [1944], U.S. War Dept., p. 321.)

Figure 1.3: First-Day Japanese Sketch (Possibly from Quarry)
(Source: "Western Pacific Operations," Hist. of Marine Corps Operations in WWII, by Garand and Strowbridge, Hist. Branch, G-3 Div., HQMC, n.d., p. 511, USA SC 301128.)

Almost every square yard of the tiny volcanic hellhole had been surveyed for indirect fire, and underground telephone cables guaranteed its adjustment.[45] Providentially, four big guns overlooking the beaches had given away their location to U.S. ships prior to D-Day. (See Figure 1.3.) And, shortly after the landing, a big underground cable was stumbled upon and severed near Suribachi.[46]

The Opposition Quantified

By the end of May 1944, according to an intercepted Japanese report, 14 coastal defense guns of 120mm or larger, 12 heavy anti-aircraft guns, and 30 25mm twin-mounted anti-aircraft guns were being operated by Iwo's Naval Guard Force.[47] Many of those weapons could be aimed directly at the beaches and isthmus from nearby high ground. The number of naval weapons almost certainly increased as the U.S. invasion approached. Additional ground units brought their own.

The enemy army personnel on Iwo at the time of the U.S. landing were as follows: (1) 640 of the 109th Infantry Division headquarters; (2) 9867 of the 2nd Independent Mixed Brigade; and (3) 3230 of the 145th Infantry Regiment. The Japanese Army was 13,737. The enemy naval personnel were: (1) 4683 in the Naval Guard Force; (2) 1425 from Naval Construction Units; and (3) 845 of Air Defense Units. The Japanese Navy total was 8038. That makes 21,775 in all according to the G-2 of TF 56.[48] Most were well armed and had plenty of ammunition. Lt.Gen. Kuribayashi was in overall control of the defenses on Iwo. He also commanded the Ogasawara Army Group and 109th Division. M.Gen. Senda was the leader of the Mixed Brigade. And Col. Ikeda commanded the crack 145th Infantry Regiment.

The Battle's Final Outcome

Kuribayashi's new defense plan would be largely successful. "By . . . [4 March], the assault elements of the 3rd, 4th, and 5th Marine Divisions were exhausted, their combat efficiency reduced to dangerously low levels,"[49] according to an official U.S. Marine Corps (USMC) battle chronicle.

A decisive victory on Iwo would take more resolve than American planners had thought necessary. Because of the required number of bunker take-downs, that resolve would come mostly from U.S. troops. Despite little chance of surviving, most had given the effort everything they had. Such an unimaginable sacrifice by that many strong young Yanks must never be forgotten. The tactical wisdom that they had collectively generated must be allowed to shape how this nation fights all future wars. The Japanese are not the only Oriental army to have mastered sub-surface combat. The Vietnamese, Chinese, and North Koreans have all become quite adept at it. So, U.S. forces may once again find themselves in an Iwo-Jima-like nightmare. Without the lessons so dearly earned on Iwo, many of those modern-day gladiators will unnecessarily die.

On that tiny WWII island, over 6100 courageous Americans would be killed in action or die of their wounds and another 17,900 be wounded. That's the only time in U.S. history where the number of friendly casualties has exceeded the number of people fought.[50] But, because of the extraordinary adversity involved, their victory still constitutes a monumental achievement by this country's greatest generation. Those "prides" of Morgantown and Flatbush had grown up during the Depression, learned to survive alone in the woods or streets, and then been trained by a military organization that had been allowed to establish its own culture. The question now is whether a modern-day American force, with all its crutch-like paraphernalia and contemporary societal mores, could do as well against the same kind of opponent.

Key Areas of Investigation

Several truths are now self evident with regard to the battle for Iwo Jima. It was not the job of Western historians to point them out. And, America's military planners have generally lacked sufficient knowledge of light-infantry tactics and the Asian thought process to fully interpret any Eastern engagement. But, no one should now worry that further knowledge about the Battle for Iwo Jima might bring discredit upon those who led or fought there. If all-out warfare lasts another thousand years, few accomplishments will be more noteworthy than how U.S. forces' were able to overcome such daunting adversity.

No amount of additional bombardment on Iwo's northern plateau would have made much difference, because of the extent to which its defenders were dug-in, camouflaged and sub-surface mobile. The island could not have been blockaded into submission, because of its water supply and speed with which the Allies needed to bomb the home islands. Nor could Iwo have been bypassed. So, only yet to be answered is what became of all Japanese defenders and the U.S. lessons learned.

The Japanese were not on Iwo, they were in it. The only way to defeat them was a blanket of all captured territory by tiny infantry contingents with enough authority to quell rear-area disturbances. This type of thing does not come easily to a top-down war machine. The requisite level of fireteam and rifleman initiative flies squarely in the face of Western military protocol. How the Marines on

Iwo managed to so quickly develop it is their gift to future U.S. warfighting. And all those involved—of whatever rank—now deserve not only this nation's undying gratitude, but also its full acceptance of their tactical insights.

2 Extent of the Japanese Excavation

- How many foe positions were linked by tunnel on Iwo?
- Was the enemy ever seen maneuvering above ground?

Japanese soldiers could stand upright in many tunnels.

(Source: PC-2600-6421-94, Washington, D.C.: U.S. Defense Intelligence Agency, 1994, p. 3-125.)

What Made Iwo Jima Different from Other U.S. Objectives

For the Battles of Peleliu, Iwo Jima, Okinawa, and Luzon, the Japanese had altered their normal defense plan. They would no longer contest the beachheads or run massive *banzai* attacks. (Similar tactics had already been proven to work by the Eastern armies of Europe.[1]) The replacement stratagems would make all U.S. advances so costly as to sap the combat power of participating units. The Japanese called it "retreat combat."[2] On Iwo, they would execute an elastic defense in depth. It would be mostly conducted through three successive belts of platoon-sized strongpoints.

Within each separate strongpoint, several Japanese machinegunners and their protective riflemen encircled a cave, excavation, or pillbox (in which to escape bombardment). All had a defiladed way to reach that opening in the ground. Common throughout the earlier island campaigns had been several automatic-weapon emplacements linked by trench to a central blockhouse. (See Figures 2.1 and 2.2.) The Iwo ridgeline variation was a tunnel through which the machinegun/rifle squads of each tiny fortress could withdraw to reverse-slope positions. When about to be overrun, those squads could either defend again from a fallback position or shoot passing adversaries in the back. That's because there was a better concealed semi-circle of automatic-weapons positions at the far end of their tunnel. (See Figure 2.3.) Just north of Airfield No. 2, lay a series of wild, rocky ridges that twisted from shore to shore.[3] They were to provide the backbone for all three belt-like Japanese defense lines.

Other strongpoints on Iwo were like those in Figures 2.4 and 2.5. As with the ridgeline fortresses, most were interconnected below ground. To move all the way back to the next defense line, their occupants had only to follow sideways tunnels to a perpendicular conduit heading toward the island's northeast end.

Such an in-depth protective array had a built-in offensive capability. Those manning any rearward portion could use existing passageways to move forward into ground already taken by the Americans. Most of its tiny portals were so well camouflaged as to be virtually invisible. Some were covered by live vegetation.[4] Others were hidden among the rocks. Still others led to concealed trapdoors in the floors of formerly occupied bunkers.

How the Marines First Chose to Deal with This Problem

It would take Marine leaders a long time to realize how mobile the foe actually was below ground. They had faced sub-surface defenses before and, on occasion, poured burning diesel fuel into them. But on Iwo Jima, this did little good because of the extent of the enemy's sideways digging and ventilation shafts. So, the Americans attempted to close every opening with Trinitrotoluene (TNT). This too met with limited success, because rock softness had made new exits easy to create. So, something else had to be done about the regularly vanishing foe machinegunners. U.S. troops would have

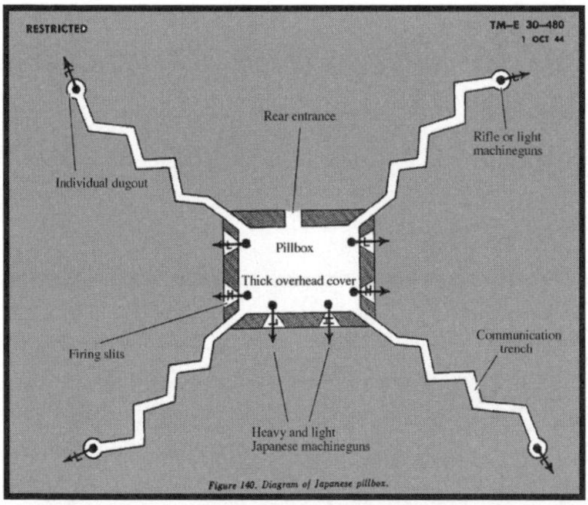

Figure 2.1: Pillbox Protection for Several Squad Strongpoints
(Source: "TM-E 30-480" [1944], p. 160.)

Figure 2.2: Iwo's Cement Pillboxes Were Stronger Than This One
(Source: TM-E 30-480 [1944], p. 161.)

advised "hunter-killer'" teams. Before long, many unit advances were being preceded by a crawling pair of low-ranking volunteers with a bag of grenades.

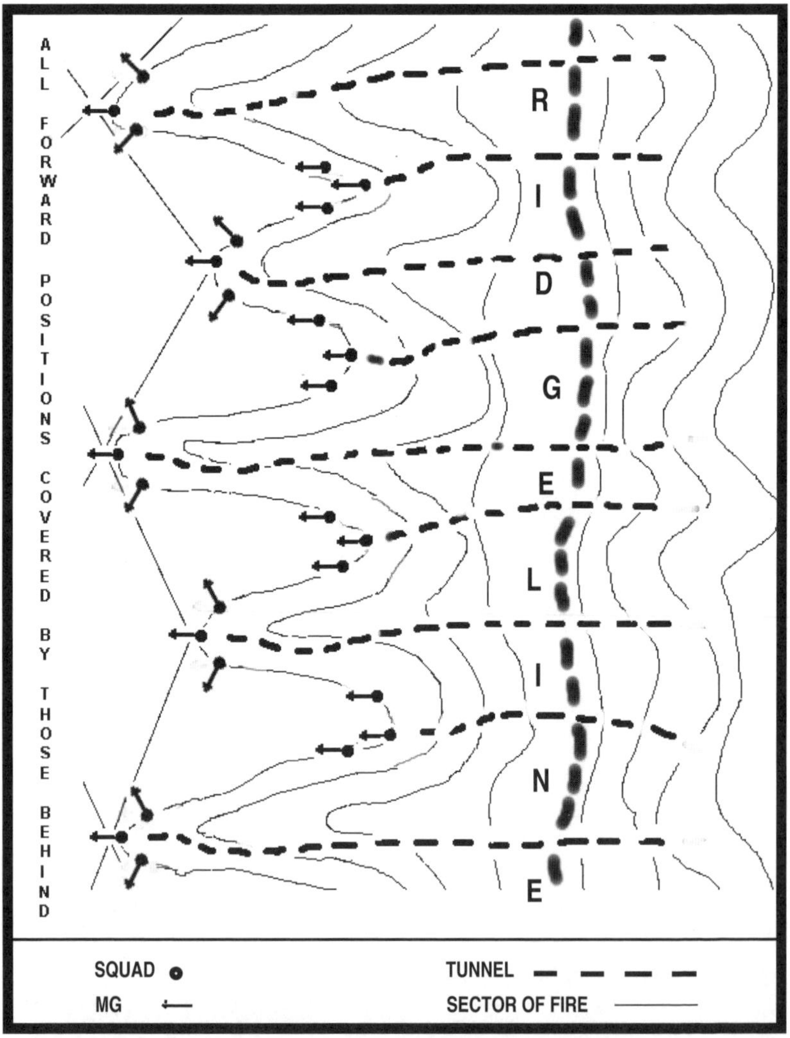

Figure 2.3: Tunnels Linked Platoon Matrices to Reverse Slope

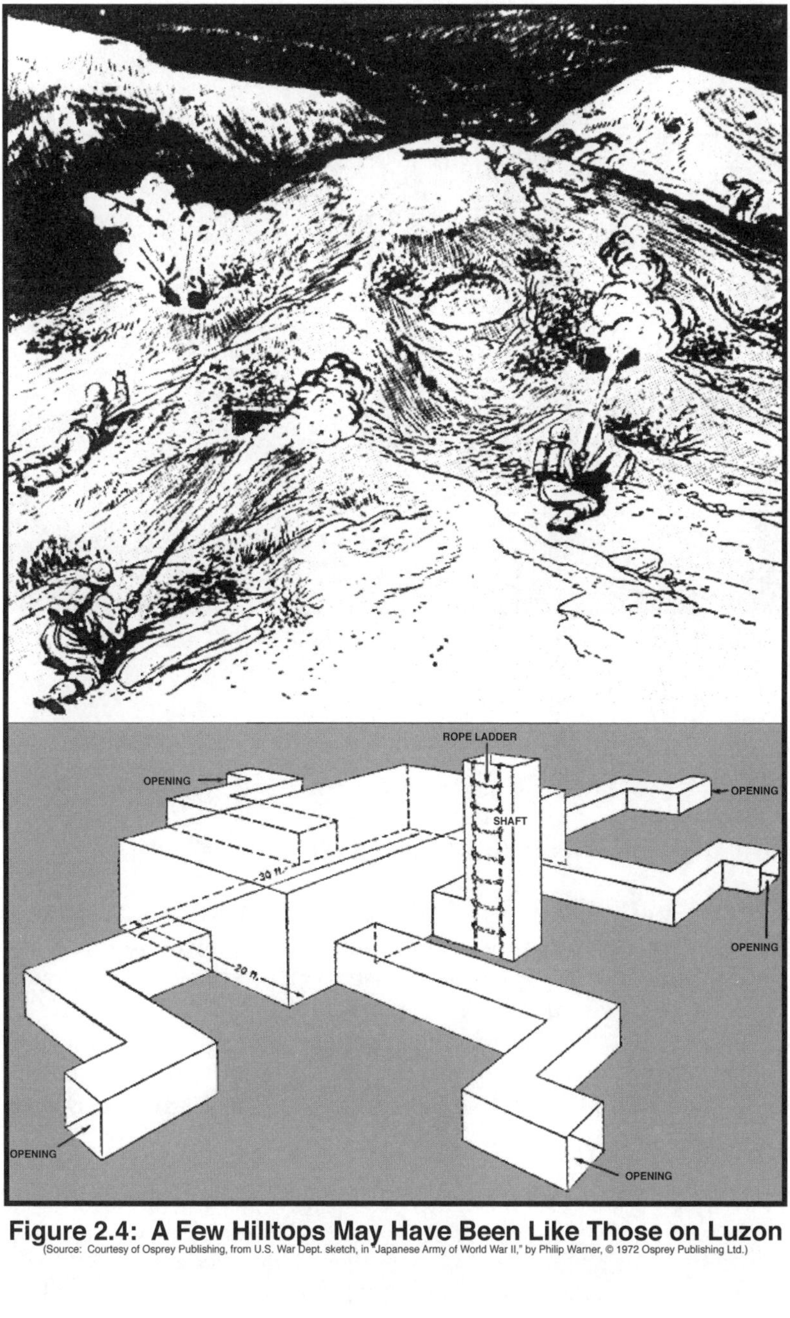

Figure 2.4: A Few Hilltops May Have Been Like Those on Luzon
(Source: Courtesy of Osprey Publishing, from U.S. War Dept. sketch, in "Japanese Army of World War II," by Philip Warner, © 1972 Osprey Publishing Ltd.)

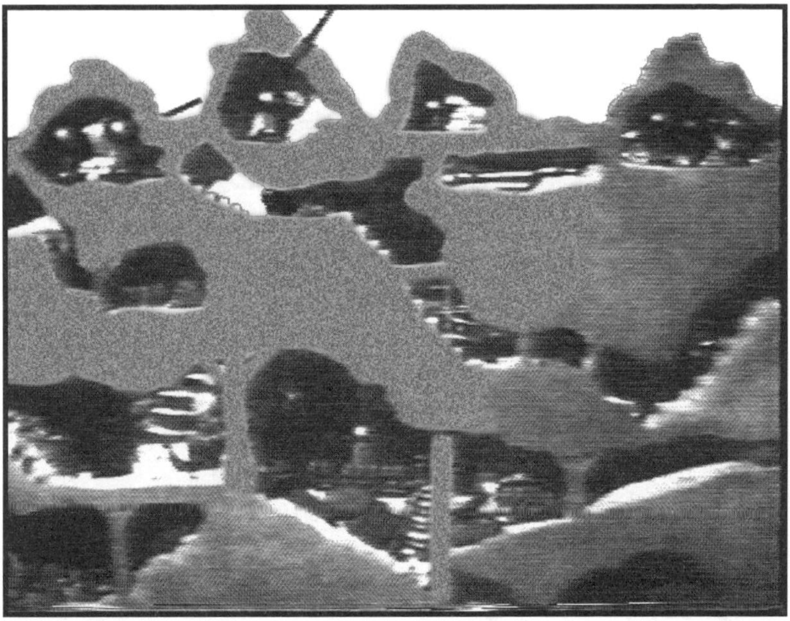

Figure 2.5: On Fairly Level Ground Were Hollow Hummocks
(Source: "A Tribute to WWII Combat Cameramen of Japan," Nippon TV, videocassette.)

How Big Was This Below-Ground Metropolis?

One still wonders how far those tunnel networks may have stretched on Iwo Jima, because all known entrances were immediately sealed. And, only the first sub-surface tier of a few major bastions were ever inspected (after all fighting had ended). In 1968, the island was given back to the Japanese. Yet, there is now every indication that every platoon-sized strongpoint along all three defense "lines" on Map 1.1 had been linked below ground before Iwo was finally declared secure by the Americans.

> The miles of tunnels dug by the Japanese, some sinking six levels deep, criss-crossed the [entire] island underground, making it hard for the Marines to pinpoint their enemy.[5]
> – *New York Times,* 15 March 1995

Details of the Tunneling Effort

Below is a list of underground facilities at Suribachi and every terrain feature near the defense lines depicted in Map 1.1.

Suribachi: Covered with gun positions, observation sites, and tunnels. 70 concrete blockhouses at its base and 50 more on its lower sides. Above them many levels of caves.[6]

1st Defense Line (north to south)
Nishi Village
West end of Airfield No. 2
The Quarry: Coast defense guns emplaced along its rim. Cliffs honeycombed with layer upon layer of machinegun emplacements.[7]

2nd Defense Line (Main Line of Resistance-MLR) (north to south)
Hill 215: Possibly fortified, 500 yards northeast of Nishi.
Nishi Ridge: Some 100 camouflaged cave entrances interconnected by tunnels, one 1000 feet long.[8]
Hill 362-A: Caves on north face; honeycombed with passageways; linked by a 300-yard tunnel with Nishi Ridge.[9] (See Figures 2.6 and 2.7.)
Hill 199-Oboe: Hollowed-out naval gun cavities and reverse-slope positions.
Hill Peter: Reverse slope covered with big-gun caves.
East end of Airfield No. 2
Hill 382: Top of hill hollowed out for several concrete artillery and antitank gun housings—each protected by as many as 10 machinegun nests—and rest of hill laced with tunnels;[10] elaborate underground honeycomb—nearly 1000 yards of caves for resupply and conduits for troops to reinforce threatened positions or to escape when one was overrun;[11] sub-surface passageways led into defenses and when one occupant got killed, another could take his place.[12]
The Amphitheater: An extension of Charlie-Dog Ridge. (See Map 2.1.) It had three levels of blockhouses connected by 700 yards of tunnels.[13]

Turkey Knob: One cave, out of hundreds designed to be defended in depth, had a tunnel 800 feet long with 14 separate exits. At the hill's top was a seemingly impregnable concrete blockhouse. And there were circular tunnels capable of sheltering 2,000 troops.[14]

Minami Village

Likely Top to the 3rd Defense Line (north to south)

Kitano Point: Entire northern tip of island was honeycombed with caves and passageways.[15]

Hill 165

Gorge and Kuribayashi's Cave: 8,500 tons of explosives were needed to detonate huge igloo-shaped concrete installation on nearby knoll. It was surrounded by mutually supporting caves.[16]

Hill 291: Had just mentioned communications center on top and other headquarters chambers of 109th Infantry Div. inside. Hill was 400-500 yards east-northeast of Kita along the edge of Gorge's feeder draw.[17]

Initial Approximation of 3rd Defense Line (north to south)

Kita Village: Complex of pillboxes and interconnected caves with communication center in big blockhouse just south of Kita (room 150 feet long and 70 feet wide). Its exit was 500-foot tunnel about 75 feet deep that led to an opening between small hills.[18]

Hill 362-B: Dense network of interconnecting caves and pillboxes.[19] (See Map. 2.1.)

Hill 357

Hill 331

Hill 362-C: Possibly linked by tunnel to caves in maze of sandstone buttes 500 yards to west.

Higashi Village

Mysterious Northeastern End of the Island

Ginmeisui 1000 yards north-northeast of Hill 362-C
Two-terrace staircase leading down to shoreline cliff.

Figure 2.6: What the Marines Claim to Be Hill 362-A
(Source: "Iwo Jima: Amphibious Epic," by Lt.Col. Whitman S. Bartley, from sketches by 31st Naval Construction Battalion, Hist. Branch, HQMC, p. 130.)

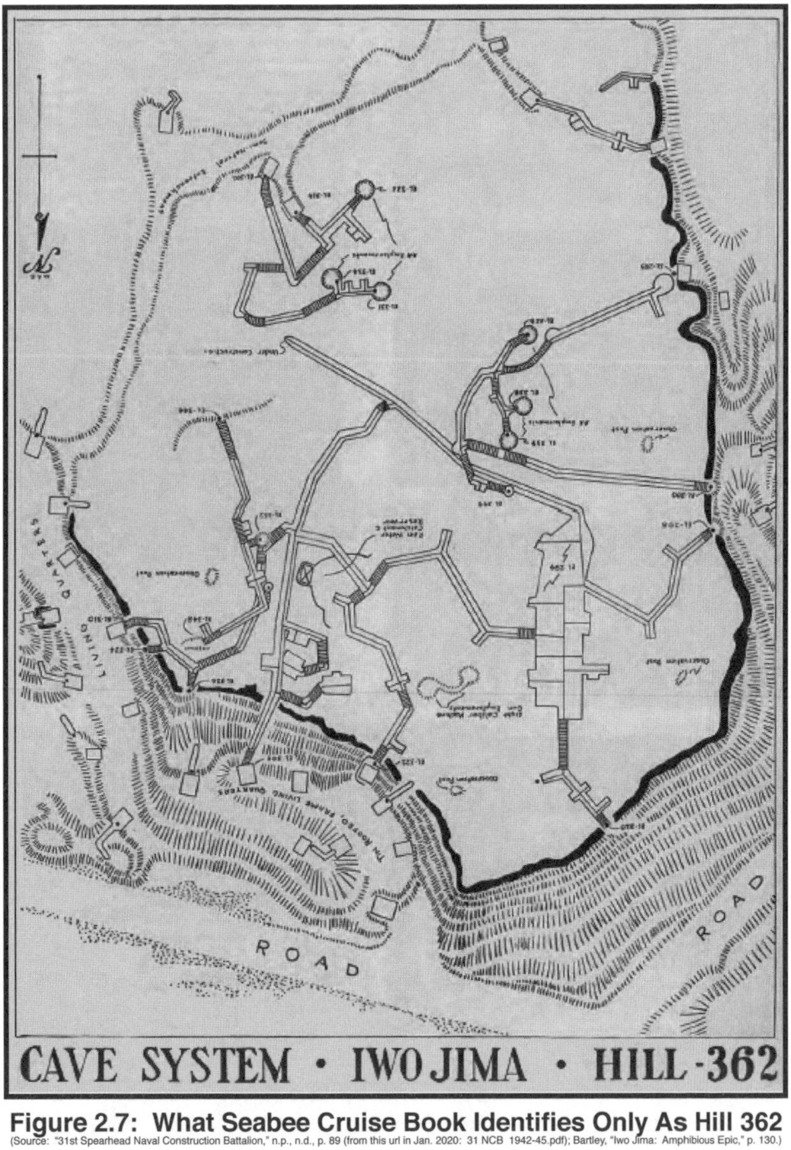

Figure 2.7: What Seabee Cruise Book Identifies Only As Hill 362
(Source: "31st Spearhead Naval Construction Battalion," n.p., n.d., p. 89 (from this url in Jan. 2020: 31 NCB 1942-45.pdf); Bartley, "Iwo Jima: Amphibious Epic," p. 130.)

A Japanese Documentary's Amazing Disclosure of 1995

In 1995, a cross-section of Iwo Jima appeared in a Japanese television special about wartime cameramen. For the first time, it showed that everything from Airfield No. 1 to just short of the north-eastern shore had been connected by underground passageway.[20] And, from Airfield No. 2 onward, there was a second larger tunnel beneath the first. It ran through a series of massive chambers. Somewhere beyond Hill 362-C (but before reaching the coast), that below ground labyrinth spread out into four separate levels. That's where the documentary's sub-surface expose prematurely ended, leaving the viewer to wonder what had existed at the shoreline itself. (See Figures 2.8 and 2.9.)

The makers of this documentary had apparently bisected Iwo along the same route that a below-ground island-long thoroughfare had followed just east of Hill 382 (and not far from Gen. Senda's initial headquarters).[21]

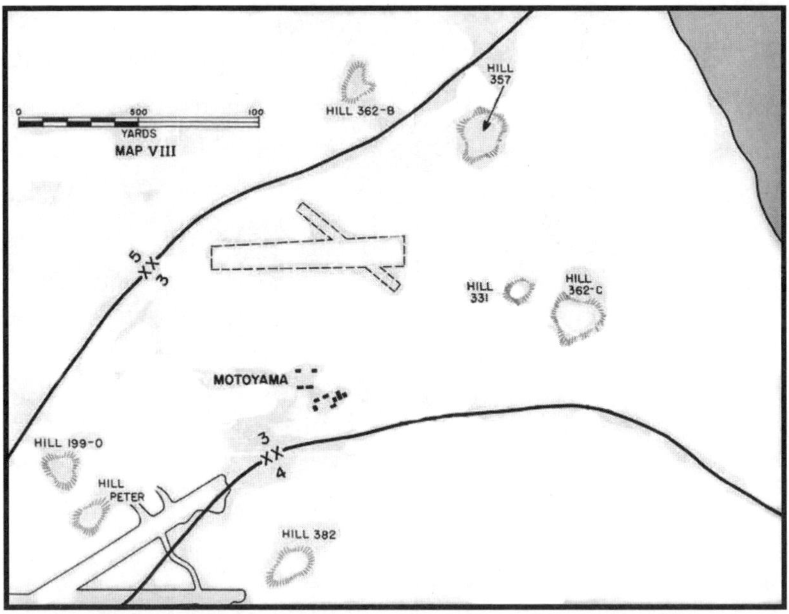

Map 2.1: Key Hills Along Opposition Defense Lines
(Source: Lt.Col. Whitman S. Bartley, "Iwo Jima: Amphibious Epic," Washington, D.C.: History Branch, HQMC, 1954, map VIII.)

A slightly modified version of Figure 2.9 appeared in the U.S. Navy's 2017 "remake" of its 1945 movie, "To the Shores of Iwo Jima."[22] But, much had been lost in the translation. The island's halves had been slid farther apart laterally, the missing northeastern end imagined, and the huge underground rooms near that end made smaller. (Within the bibliography are internet locations of both movie editions). Closely comparing the Navy's new vision with a historically correct scale model confirms major changes to the island's shape. (See Figure 2.10.) When properly aligned (as in Figures 2.8 and 2.9), the rear slice should stick farther out into the ocean than the front slice. The original movie had shown no cross-section whatsoever, or any of the other introductory material. There is no evidence of American service personnel ever fully inspecting Iwo's tunnel system, so the U.S. Navy should not now be claiming to have known its particulars in 1945.

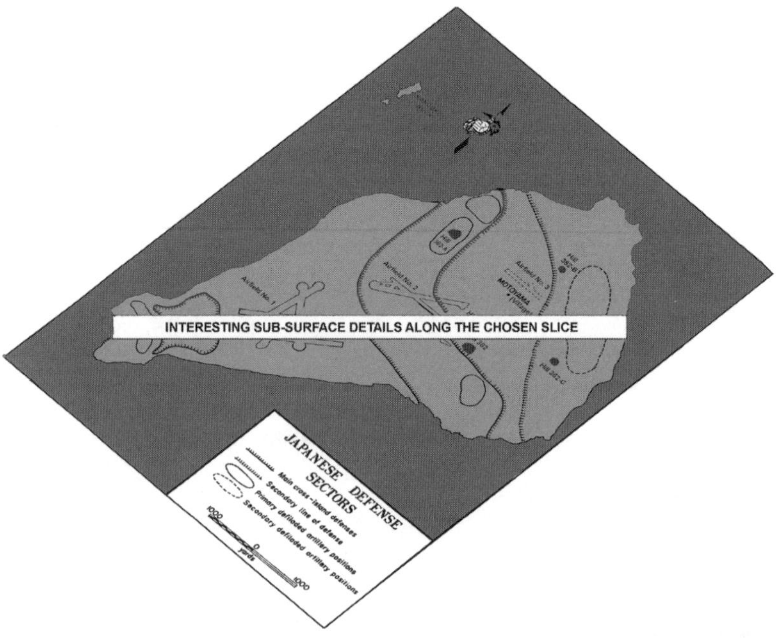

Figure 2.8: Where Japanese Documentary Had Split Island
(Source: "A Tribute to WWII Combat Cameramen of Japan," Nippon TV videocassette; "Closing In: Marines in the Seizure of Iwo Jima," Hist. & Museums Div., HQMC, p. 8.)

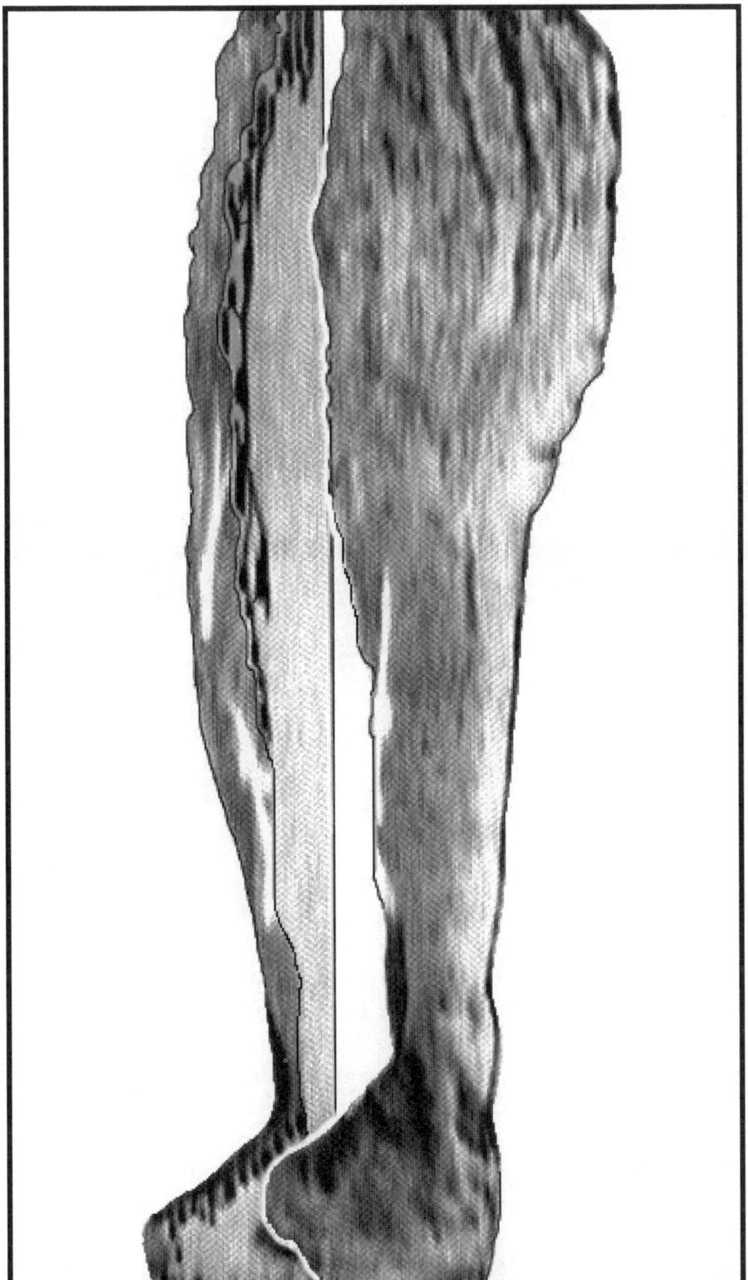

Figure 2.9: Cross-Section Showing Sub-Surface Rooms and Several Levels to Island-Long Tunnel
(Source: "A Tribute to WWII Combat Cameramen of Japan," Nippon TV, n.d., videocassette; "Reflections on Iwo Jima: The Japanese Tunnels," Youtube presentation, Marines TV, 26 March 2011.)

Figure 2.10: Iwo Jima's Northeastern End
(Source: Courtesy of Howard Models of Toledo, OH, image designator "IwoJima2s.jpg," copyright © Howard Models.)

The next frame of that Japanese special (and remade Navy movie) resembles a glassed-in anthill. (Look back at Figure 2.5.) It reveals underground details of a flatland matrix on Iwo: (1) surface hummocks hollowed out as gun emplacements; (2) several levels of subterranean rooms; and (3) an interconnecting tunnel network with many surface portals. Some of the lower spaces appear to have ammunition and supplies hidden beneath their floors.[23] Such a multi-level arrangement may have been less prevalent on the low-ground isthmus than the central plateau. That would help to explain why the island had been so easily crossed on the first day of the invasion.

Tunneling Statistics

Kuribayashi had planned to connect his various positions with 38,000 meters of underground passageways.[24] By the time of the invasion, three reputable sources report 26,000 meters completed.[25] This would have been enough to link most beltway bastions. The combined length of the three defense lines was only 16,000 meters. The distance between Airfield No. 1 and the island's northeastern shore was another 5000. Only one source has the most difficult excavation—the 3000 meters between Airfield No. 1 and Mount Suribachi—ever being completed.

> The tunnel system was so extensive that many of the troops that were defending Mount Suribachi managed to escape to the north before the volcano fell.[26]
> — *History of War*

To fully execute this new style of defense, Kuribayashi had gotten professional construction advice and a multitude of forced laborers from Korea.

> Kuribayashi demanded the assistance of the finest mining engineers and fortifications specialists in the Empire. Here again the island favored the defender. Iwo's volcanic sand mixed readily with cement to produce superior concrete for installations; the soft rock lent itself to rapid digging. Half the garrison lay aside their weapons to labor with pick and spade. When American heavy bombers from the Seventh Air Force commenced a daily pounding of the island in early December 1944, Kuribayashi simply moved everything—weapons, command posts, barracks, and stations—underground. These engineering accomplishments were remarkable. Masked gun positions provided interlocking fields of fire, miles of tunnels linked key defensive positions, every cave featured multiple outlets and ventilation tubes. One installation inside Mount Suribachi ran seven stories deep. The Americans would rarely see a live Japanese on Iwo Jima until the bitter end.[27]
> — *Closing In: Marines in the Seizure of Iwo Jima*
> History and Museums Division, HQMC

Precisely how much connector tunnel had been finished by D-Day may never be known. But one definitely ran the full width of the second-defense-line MLR. Nishi Ridge had been connected to Hill 362-A. And Maj.Gen. Senda had safely moved his headquarters from just east of Hill 382 to the 4th Division's final pocket—some 1700 meters to the southeast (more on this in Chapter 10).[28] The fighting had been so widespread at the time that most of this journey must have occurred below ground.

It's now known that every piece of key terrain on the island already had its own honeycomb of underground passages, and that Japanese tunnelers could dig at a rate of 1/3 meter per hour.[29] So, during the 36-day battle, a single team of diggers could have advanced 300 meters. The 12,000 meters still missing from Kuribayashi's stated goal would have required only 40 teams.

Because of the importance of the underground network to defense of the island, 25% of the garrison may still have been tunneling after the invasion began. That's 5000 Japanese soldiers helping up to a thousand forced laborers from Korea. Protective spaces below ground ranged in size from small caves for a few men to chambers big enough to hold 300 or 400. Further to prevent Japanese combatants from becoming trapped, most subterranean installations had multiple levels and exits.[30]

What Kuribayashi's "Last Stand" Had Suggested

The extent of tunneling between the center of the 1st Defense Line and top of the 3rd didn't become fully apparent until the night of 25 March, two weeks after the island had been "declared occupied" through a second flag-raising.[31] That's when some 300 Japanese regulars from a fully surrounded Gorge supposedly attacked the billeting area of Airfield No. 1.[32] Such a feat is almost impossible to perform through infiltration alone. POWs (prisoners of war) would later claim that Kuribayashi had personally led the excursion.[33]

> General Schmidt received the good news that the 5th Marine Division had snuffed out the final enemy cave in The Gorge on the evening of D+34 (25 March). But even as the corps commander prepared his announcement declaring the end of organized resistance on Iwo Jima, a very well-organized enemy force emerged from northern caves

> and infiltrated down the length of the island. This final spasm of Japanese opposition still reflected the influence of Kuribayashi's tactical discipline. The 300-man force took all night to move into position around the island's now vulnerable rear base area, the tents occupied by freshly arrived Army pilots of VII Fighter Command. . . .
>
> . . . Sunrise revealed . . . 300 dead Japanese; more than 100 slain pilots, Seabees, and pioneers; and another 200 American wounded. It was a grotesque closing chapter to five continuous weeks of savagery.[34]
>
> – *Closing In: Marines in the Seizure of Iwo Jima*
> History and Museums Division, HQMC

Tunnels Had Likely Been Used for That Final Raid

Could the beleaguered defenders of Kuribayashi's cave have done all of the following in the same night: (1) exfiltrate the Marine cordon; (2) sneak three miles down the island; and (3) then infiltrate the Airfield No. 1 perimeter? Or might some other Japanese element nearer the airfield have taken advantage of existing passageways? Sub-surface facilities often have hidden stairs to lower levels. So, even if U.S. forces had done some searching below ground on their way through Iwo's plateau region, they could have easily bypassed enemy contingents.

On the way up the Gorge from its southern "headwaters," attacking U.S. units had collapsed a huge blockhouse and then sealed the most obvious entrances to Kuribayashi's Cave.[35] What then occurred inside that Cave may never be known. It was later inspected by the Seabees, but their sketch made no mention of an outbound tunnel. (See Figure 2.11.) On 4 June, a patrol from Company F of the Army's 147th Infantry Regiment found people still in what had probably been the same headquarters (though its location was only reported as "in the island's northeast quadrant"[36]). It could only have been remanned through a Third-Defense-Line tunnel.

Kuribayashi Had Planned a Guerrilla War

As at Peleliu, Iwo's new style of defense led to guerrilla warfare. Repeated raiding such a heavily occupied space takes tunnels.

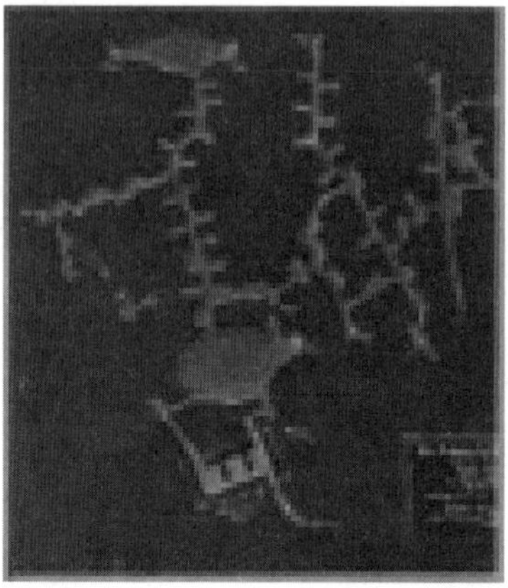

Figure 2.11: One-Third of Kuribayashi's Multi-Level Gorge Cave
(Source: 31st [Naval Construction Battalion] Spearhead," n.p., n.d., p. 89, as retrieved in February 2020 from the following url: 31 NCB 1942-45.pdf.)

> Kuribayashi therefore decided . . . to conduct all defenses below ground, linking them with tunnels so that the defenders could move between them. . . .
>
> . . . Kuribayashi felt the hiding in their underground positions and making sneak attacks when the enemy was least expecting them was the most effective option. With such tactics, the Japanese could keep on fighting when down to their last man. It was guerrilla warfare.[37]
>
> — Compilation of Kuribayashi's letters

Enduring Speculation

Did the missing 10,000 defenders all run out of oxygen in those sealed-up passageways? After all Marines had departed the island, the 147th Infantry Regiment found 2500 alive.

> The 9th Marines . . . became the last to leave, conducting two more weeks of ambushes and combat patrols. The 147th Infantry inherited more of the same. In the first two months after the Marines left, the Army troops killed 1,602 Japanese and captured 867 more.[38]
>
> — *Closing In: Marines in the Seizure of Iwo Jima*
> History and Museums Division, HQMC

And those American soldiers additionally encountered many enemy troops with plenty of food and water. The assumption at the time was that they had stolen it.

> Many Japanese, whether dead or captured, showed no signs of starvation or privation at all.[39]
>
> — "The 147th Infantry Regiment at Iwo Jima"
> U.S. Army Center of Military History

The ultimate source of all that food and water is still in doubt. Once the U.S. Navy blockade had been lifted, Japanese submarines could have possibly reached the island. If there were still wounded combatants in the huge underground chambers at its northeastern end, they may have been minimally resupplied. Japanese holdouts continued to emerge from Iwo Jima's rubble until January 1949.[40] (Look back at Map 2.2 for where Japanese units had been originally stationed.)

Back to That Island-Long Passageway

The Japanese documentary cross-section had reached the northeastern coast of Iwo Jima around Grid Coordinate (GC) 806760 (just behind Hill 123). (Look back at Figure 2.8 and then at Map 2.3.) Just south of its end was an elongated islet some 50 yards offshore and then a machinegun-infested promontory.[41] A post-bombardment aerial photograph shows this same coastal area being disturbed, either by Japanese construction or U.S. bombing.[42] A pre-invasion map depicts back-to-back trenches there as well. The only other spots on Iwo to have such trenches were its isthmus beaches.[43] Along this remote part of Iwo's northeast coast had also been strung the only barbed wire on the island.[44] It was being actively defended, but why?

Where the documentary tunnel led was obviously important, but battlefield events had brought little attention to it. Kuribayashi's radio/telegraph transmissions back to Japan had instead highlighted

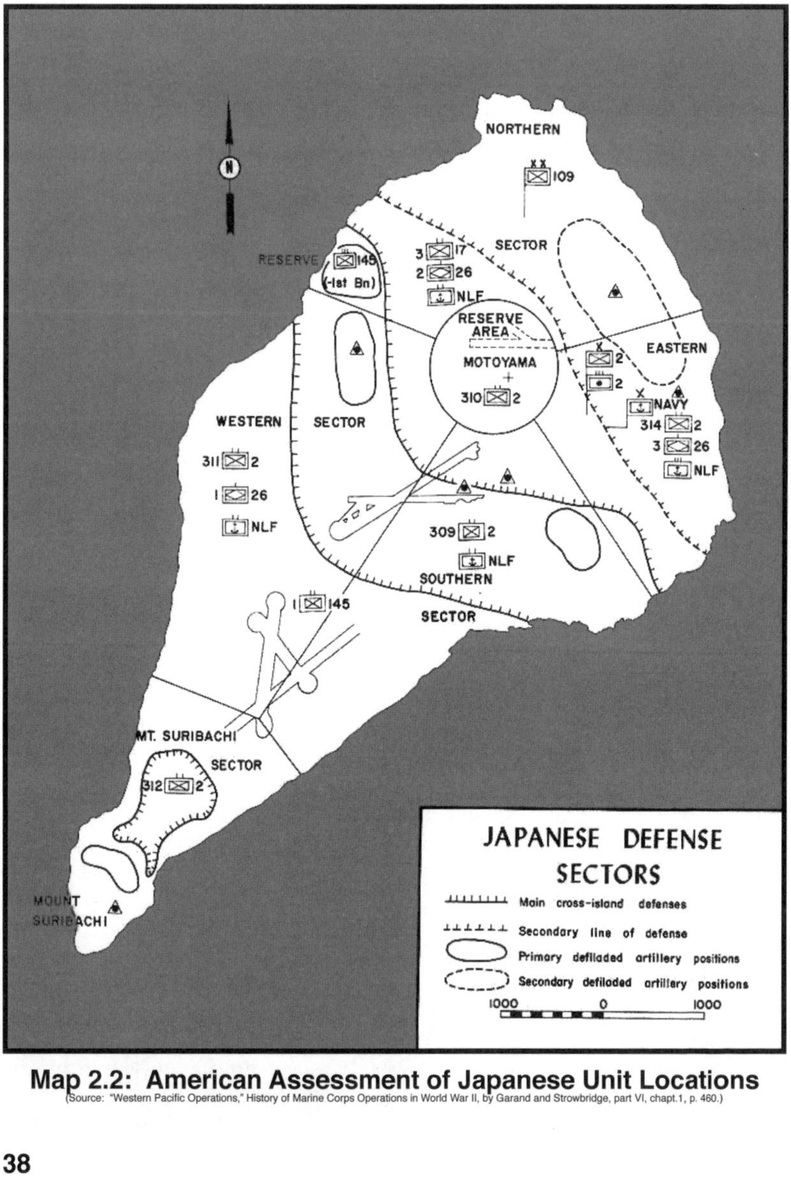

Map 2.2: American Assessment of Japanese Unit Locations
(Source: "Western Pacific Operations," History of Marine Corps Operations in World War II, by Garand and Strowbridge, part VI, chapt.1, p. 460.)

the Gorge. If the last stand for all Iwo defenders was to be the island's northern sector, why did the Third Defense Belt stretch all the way to Tachiiwa Point? And why was the central hill-mass in this belt—362-C—so hard to take. Like Hill 382, 362-C was quite near the documentary conduit (closer than some U.S. maps portray). (See Map 2.3.) Farther along that conduit lay a "low-ground" artillery observation post (OP) and the mysterious *Ginmeisui*. (Look back at Map 1.1.)

Nor would the Japanese have fought so hard for Cushman's Pocket (just southeast of Airfield No. 3). That "jungle of stone" may have had its own entrance to the island-long thoroughfare.

> The enemy position [at Cushman's Pocket] was a maze of caves, emplaced tanks, stone walls and trenches.[45]
>
> – *Closing In: Marines in the Seizure of Iwo Jima*
> History and Museums Division, HQMC

Once such questions are answered, there may be a more definitive explanation of what happened to the other 7500 Japanese soldiers who supposedly ran out of oxygen in the tunnels.

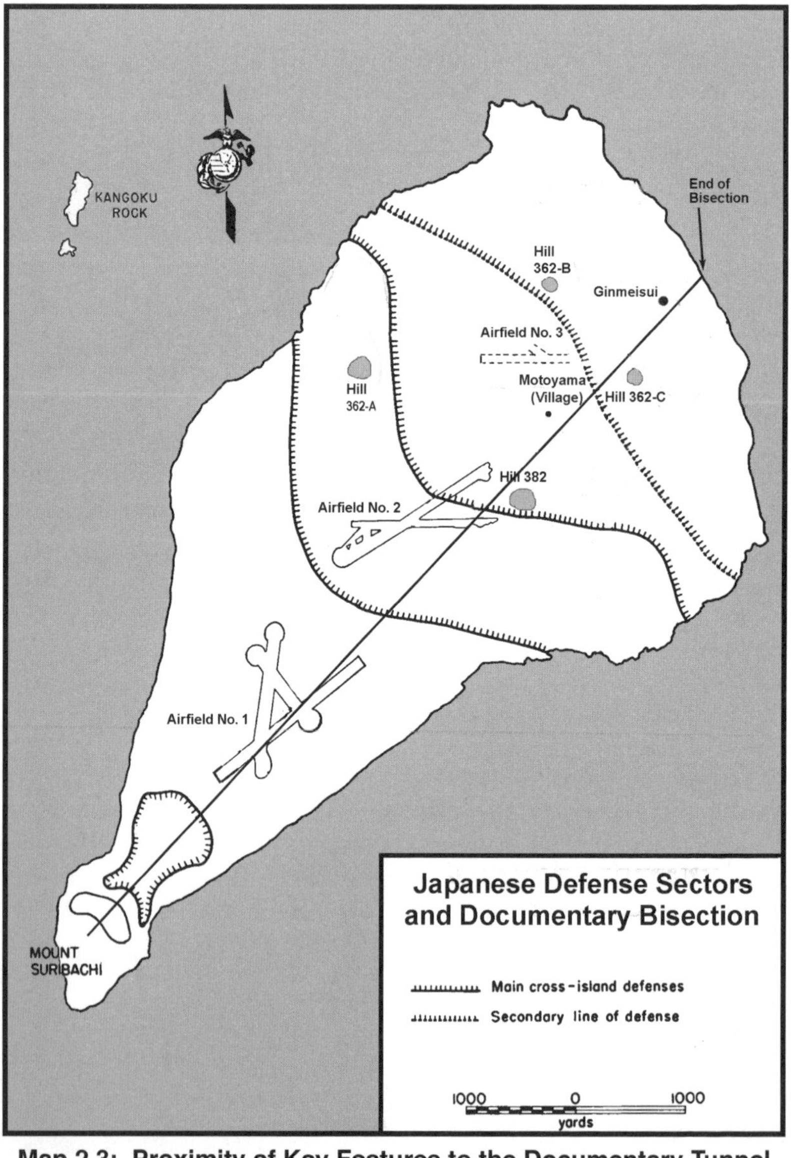

Map 2.3: Proximity of Key Features to the Documentary Tunnel
(Source: "Closing In: Marines in the Seizure of Iwo Jima," by Alexander, Hist. & Museums Div., HQMC, pp. 5, 8.)

Part Two

Tactical Issues

"Uncommon valor was a common virtue on Iwo Jima."
— Admiral Chester Nimitz

(Source: AZ Quotes, s.v. "Chester W. Nimitz," as retrieved in October 2018 from this url: https://www.azquotes.com/author/10830-Chester_W_Nimitz.)

3 A State-of-the-Art Japanese Defense

- What did U.S. forces encounter on Iwo that was unusual?
- Which challenges came as a surprise?

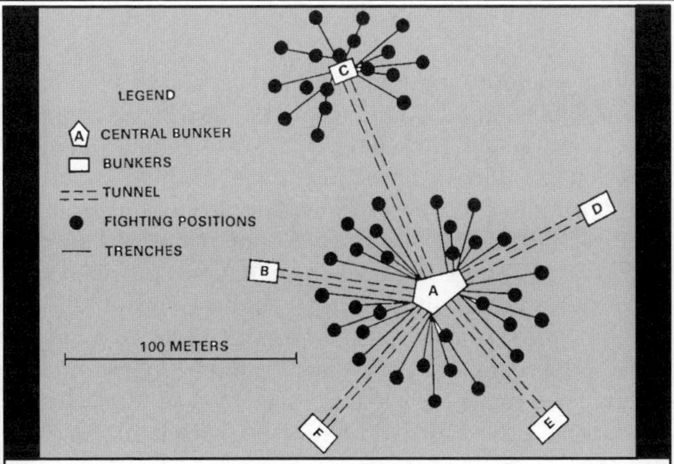

A 1970's matrix of interconnected platoon strongpoints.

(Source: "Counterguerrilla Operations," FM 90-8 [1986], p. A-6.)

What Constitutes a Tactically Sophisticated Defense

After the machinegun's invention, the first evolutionary advance to defensive tactics came from the WWI German Stormtroopers. Instead of a continuously manned trenchline, they created an in-depth squad strongpoint matrix. Then, wherever Allied troops thought they were exploiting a gap, they ran into a firesack. The next advance occurred during WWII. That belt of tiny mutually supporting forts went below ground and then mimicked the new urban defense—only to fight where the enemy was attacking and then gradually to move backwards. The next development was summarized by President Kennedy in 1962 while describing the Unconventional Warfare (UW) part of 3rd Generation Warfare (3GW)—"ambushes on defense, and infiltration in the attack."[1] 3GW involved going after a foe's command and logistics apparatus, instead of his frontline troops. What Kennedy had been talking about became perfectly clear in 1979. That's when a Chinese army with modern aircraft and tanks had trouble penetrating a flatland array of strongpoints like those at the beginning of this chapter. Because the North Vietnamese Army (NVA) had been off fighting the Khmer Rouge in Cambodia, most of the damage was done to the invading Chinese army by North Vietnamese border militia with hand-held weaponry.[2]

Iwo Jima's defense planners had almost certainly followed the same evolutionary trends. But what they produced was so lethal that it set a few tactical precedents of its own. One would likely be how to mesh a fire plan with broken ground. In the West, fairly level terrain is normally chosen for a blocking formation, as it permits interlocking bands of grazing fire. Yet, at Iwo, the Japanese had preferred to defend jumbled real estate, because it provided a natural barrier to flame tanks and artillery shells. They then directed most machinegun fire down draws that could serve as opposition avenues of approach, and covered with mines or indirect fire all depressions that could harbor enemy troops. A quick look at all Japanese operations on Iwo Jima may bring to light other tactical innovations.

Still Active Blockhouses at Center of Debarkation Zone

As the Marines crossed the second terrace above Yellow Beach, they encountered automatic weapon fire from two nearby pillboxes.[3] (Look back at the cover artwork.) Those pillboxes may have been some of the shoreline "fortifications" that had been pummeled by U.S. planes since June 1944. Their shattered remains had simply been reoccupied through tunnels.

Above Red Beach, one of the Corps most famous SNCOs—"Manila John" Basilone—encountered a similar problem.[4] Pre-invasion fire had undoubtedly damaged his pillbox as well, yet its machineguns were still active. The enemy was starting to regularly show up where he shouldn't have been.

> 19 February 1945. . . . Shrewdly gauging the tactical situation shortly after landing when his company's advance was held up by the concentrated fire of a heavily fortified Japanese blockhouse, Gunnery Sergeant Basilone [of C/1/27] boldly defied the smashing bombardment of heavy caliber fire to work his way around the flank and up to a position directly on top of the blockhouse and then, attacking with grenades and demolitions, single-handedly destroyed the entire hostile strong point and its defending garrison.[5]
>
> — Navy Cross citation for John Basilone

Basilone's biography has him directing the attack of a demolitions man and flame man, and then personally moving forward to machinegun 8-10 escaping occupants. It even contains a *post-mortem* photograph of those same slightly charred enemy soldiers. In this other version of the action, the standard order of bunker steps has also been changed—demolitions came before flame.[6]

Heaviest Resistance to the Landing Occurred to the East

There were recessed machineguns surrounding the big guns sticking out the Quarry cliffs on the right end of the landing zone, and from Suribachi's steep slopes on the left end. From nearly invisible portals, automatic weapons had been taking turns "mowing down" the new arrivals.

> One of the wounded Marines we received aboard when the hospital ships were overtaxed, declared the Japanese had camouflaged trapdoors all over the face of the cliff and that they would open a door, pour out a murderous volley, and close the door again, leaving the attackers staring at an apparently blank area of mountainside.[7]
>
> — *Marine Corps Gazette,* February 2000

A battalion from the 4th Marine Division had been tasked with clearing the Boat Basin end of the zone. That would take some doing.

Blue Beach Two lay directly under the observation and fire

> of suspected Japanese positions in the Rock Quarry, whose steep cliffs overshadowed the right flank like Suribachi dominated the left. . . . 3rd Battalion, 25th Marines, took the heaviest beating of the day on the extreme right trying to scale the cliffs leading to the Rock Quarry.[8]
>
> – *Closing In: Marines in the Seizure of Iwo Jima*
> History and Museums Division, HQMC

After finally ascending those cliffs, whole Marine companies of this battalion and others were to encounter intense automatic weapons fire "coming from all over."[9]

> [I]t was the strength of character of Captain James Headley and Lieutenant Colonel "Jumping Joe" Chambers who led the survivors of the 3d Battalion, 25th Marines, onto the top of the cliffs. The battalion paid an exorbitant price for this achievement, losing 22 officers and 500 troops by nightfall.[10]
>
> – *Closing In: Marines in the Seizure of Iwo Jima*
> History and Museums Division, HQMC

Some of the worst initial fighting of the entire operation would ensue.[11] 1/24 had come to the assistance of the 25th Marines, but to little avail.

> There was no cover from enemy fire. Japs dug in reinforced concrete pillboxes laid down interlocking bands of fire that cut whole companies to ribbons. Camouflage hid all enemy positions. . . . The high ground on either side was honeycombed with layer after layer of Jap[anese] emplacements. . . . Their observation was perfect; whenever a Marine made a move, the Japs would smother the area in a murderous blanket of fire.[12]
>
> – *Closing In: Marines in the Seizure of Iwo Jima*
> History and Museums Division, HQMC

Then, something happened which is all too often part of 2nd Generation Warfare (2GW or trying to kill as many enemy soldiers as possible). There were two friendly fire incidents: (1) an airstrike hitting 11 Marines; and (2) naval gunfire that took out another 90.[13]

Suribachi Itself

All the while, the left half of the landing zone was still being pummelled by shells from Suribachi. Some were coming from the gun-studded cliffs at its base. But the caves farther up its slope had also been hard to silence. Their sub-surface connections had allowed many to come back to life.

> Suribachi's caves were as formidable as its pillboxes and blockhouses. Like most enemy defenses, they had from two to five entrances with interconnecting tunnels and had been used as air raid shelters and living quarters. Some of them burrowed 50 feet below ground with steps and air shafts leading to the lowest levels of the volcano.[14]
>
> — 5th Marine Division website

Had Foe Taken Page Out of New Urban Warfare Manual?

One should not now try—from an armchair—to take fewer casualties in the Iwo landing zone. The area at its eastern end had anchored the foe's first and second cross-island defense lines. (Refer back to Map 1.1.) Against that much fortification, enemy attrition through overwhelming firepower seemed the Marines' only option. But, the Japanese had also come up with a way to dodge that firepower, and its details should be of great interest to contemporary U.S. warfighters.

By moving into multiple sub-surface layers, Kuribayashi had created the inverted equivalent of a built-up environment. Then, like the Russians had done at Stalingrad and Germans would do at Berlin, he was able to channel most of his assets to where the enemy was attacking and then pass unused portions to the rear. By the early 1950's, this was to become the state of the art for urban defense and a recognized staple of 3GW—stopping a city takeover through ambushes in series. It would be repeated by the Japanese at Manila, North Koreans at Seoul, and North Vietnamese at Hue City. The steps of each confrontation were as follows: (1) minefield; (2) ambush; (3) barricade; (4) strongpoint matrix; and (5) counterattack. Of most interest would be what happened to the "last-stand" personnel at Berlin, Manila, and Hue City. Most were to simply vanish.[15]

At all of Iwo's beach exits, U.S. tracked vehicles encountered mines. The loose soil contained enough metallic characteristics to render standard U.S. mine detectors unreliable.[16]

> Once ashore and clear of the loose sand along the beaches, the tanks, half-tracks, and . . . bulldozers . . . ran into the strongest minefield defenses, yet encountered in the Pacific War.[17]
>
> – *Closing In: Marines in the Seizure of Iwo Jima*
> History and Museums Division, HQMC

Iwo's many tiny ridgelines would suffice as barricades in the urban sequence. Then, on the very first day of the invasion, Marines units found themselves in a 360° ambush as they approached the terrain rise at Airfield No. 1.

> Ninety minutes after landing [on D-Day], the battalions of Lieutenant Colonel John A. Butler and Major John W. Antonelli had clawed up the slope to the eastern edge [of Airfield No. 1]. Resistance was light until 10:45 [A.M.] when, as Antonelli put it, "crap hit the fan in copious quantities." Moments before, they were catching only small arms fire. Now, when Kuribayashi pulled the plug, they were in the middle of the ambush and taking a deadly fusillade of machinegun bursts and mortar fire from concealed bypassed positions in the hummocks.[18]
>
> – *Iwo Jima,* by Ross

Just to reach the isthmus center, the Marines of the 27th Regiment had wended their way through a maze of pillboxes and machinegun nests (the boundary of a defense zone). (Look back at Map 1.1.) Then, all along the airstrip terrace must have been more enemy positions. Whatever the grazing fire from all those tiny forts couldn't reach, airbursts from the anti-aircraft guns on Suribachi could.[19] But for now, there would be no fifth-step counterattack in the urban-warfare sequence.

Nearly Impregnable MLRs Do Exist

Once in a while, the landscape of a battlefield can so well mesh with its fire plan that the ground becomes nearly impossible for any intruder to enter. In effect, every square yard has been covered by either direct or indirect fire. When that same real estate is also honeycombed—from front to back—with tunnels, spoiling attacks can be launched from the intruders' rear. And then, no intruder-occupied part of the battle-space is secure. Such a thing would happen at the north end of the infamous Shuri Line on Okinawa. In fact, the Shuri Castle segment of that Line would eventually have to be taken from a flank.[20] That's how much resistance the liberators of Iwo Jima were about to face along all three defense lines.

Degree of Mutual Support Between Composite Elements

Not only were the platoon strongpoints on Iwo mutually supporting with machineguns, but so were some of the major bastions with bigger weapons. During Battle of the Meatgrinder at the center of the enemy's Second Defense Belt, the Marines would have to attack Hill 382, Turkey Knob, and the Amphitheater all at the same time to make any forward progress.[21]

The Sub-Surface Detail of a Key Bastion

A Japanese POW was to sketch the sub-surface defense matrix (just to the east of Airfield No. 2) from which he had just come. That's about where the Amphitheater was located. (See Figure 3.1.) His drawing appears to show a tiny ridgeline, linear mound, or covered ditch containing two things: (1) a perfectly rectangular grid of some 60 machinegun emplacements; and (2) and a separate string of support establishment facilities. Such a well-ordered arrangement could not have been hewn out of solid rock; it must have been constructed above ground and then buried. About 20 Japanese infantry squads (two companies with 15 men and one machinegun per squad [22]) could have manned it (while gradually moving backwards). All rows of bunkers may have been at different levels on the forward-face of a gradual incline, or half on its reverse-slope. Instead of tunnels, all gun sites were linked by covered trenches.

In the bowl-shaped Amphitheater next to Turkey Knob, the 204th (Japanese) Naval Construction Brigade is known to have constructed a nearly impenetrable defense complex. It may have contained several back-to back fortified mounds, like the one in the drawing. That's because one battle chronicle mentions "[semi-]circular tunnels capable of sheltering 2,000 troops."[23] And another talks about hundreds of "caves" arrayed in depth along lengthy tunnels.

> One cave [at the Amphitheater] had a tunnel 800 feet long with 14 separate exits; it was one of hundreds designed to be defended in depth.[24]
>
> – *Closing In: Marines in the Seizure of Iwo Jima*
> History and Museums Division, HQMC

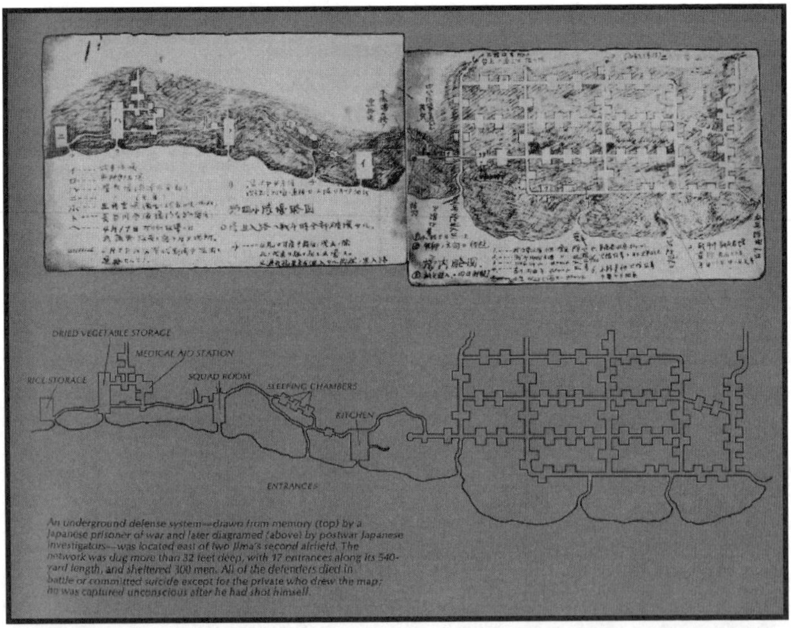

An underground defense system—drawn from memory (top) by a Japanese prisoner of war and later diagramed (above) by postwar Japanese investigators—was located east of Iwo Jima's second airfield. The network was dug more than 32 feet deep, with 17 entrances along its 540-yard length, and sheltered 300 men. All of the defenders died in battle or committed suicide except for the private who drew the map; he was captured unconscious after he had shot himself.

Figure 3.1: Sub-Surface Strongpoint Cluster at Amphitheater

(Source: "Japanese Defenses and Fortifications Tawara, Iwo Jima, and Okinawa 1943- 1945," by J.D. Rogers and K.F. Hasselmann, Univ. of Missouri at Rolla, n.d., p. 54.)

Artillery Pieces Also Protected from Bombardment

In the artillery parks on Map 1.1, many of the guns had metal doors and overhead camouflage. Most troublesome to the Marines were the 320mm spigot mortars. They ejected a rocket-assisted cylinder of explosives that was almost as large as a 55-gallon drum. One of these may have found Gy.Sgt. Basilone at Airfield No. 1, because four men were killed at the same time he was mortally wounded.[25]

The Danger of Penetrating a Japanese Defensive Array

When about to be overrun, the occupants of caves or pillboxes and their associated machinegun nests would often withdraw below ground to other emplacements.[26] Then, Japanese mortars and artillery would pound the place just vacated. After dark, those same people could move back into their former positions—to do all sorts of rear-area damage to the new owners.[27]

A Marine patrol had initially climbed to the top of Hill 382 without much resistance, but was then forced to come right back down. It had just spotted a bypassed string of pillboxes that made the hilltop too dangerous to occupy.[28]

Encirclement and annihilation have long been an integral part of Eastern strategy. Anywhere within WWII Europe, the breach of a "soft" strongpoint belt could be quickly closed to entrap the penetrating unit.[29]

Normally Required to Make Any U.S. Advance

While several large Marine units were finally forced to attack all three terrain features at the center of the Second Defense Belt in concert, their members did not all move forward at once. Most provided covering fire, while a few brave souls advanced. Company-sized assault lines had proven so costly above the Rock Quarry on the first day ashore,[30] that they had mostly ceased to occur under heavy fire.

How those heroic individuals managed to gain ground is interesting. On Hill 362-A, at the northern end of the first Japanese defense line, Cpl. Tony Stein employed a modified aircraft machinegun to get close enough to the targeted bunker for demolition placement. He and Sgt. Merritt M. Savage were to neutralize up to eight such bunkers.[31]

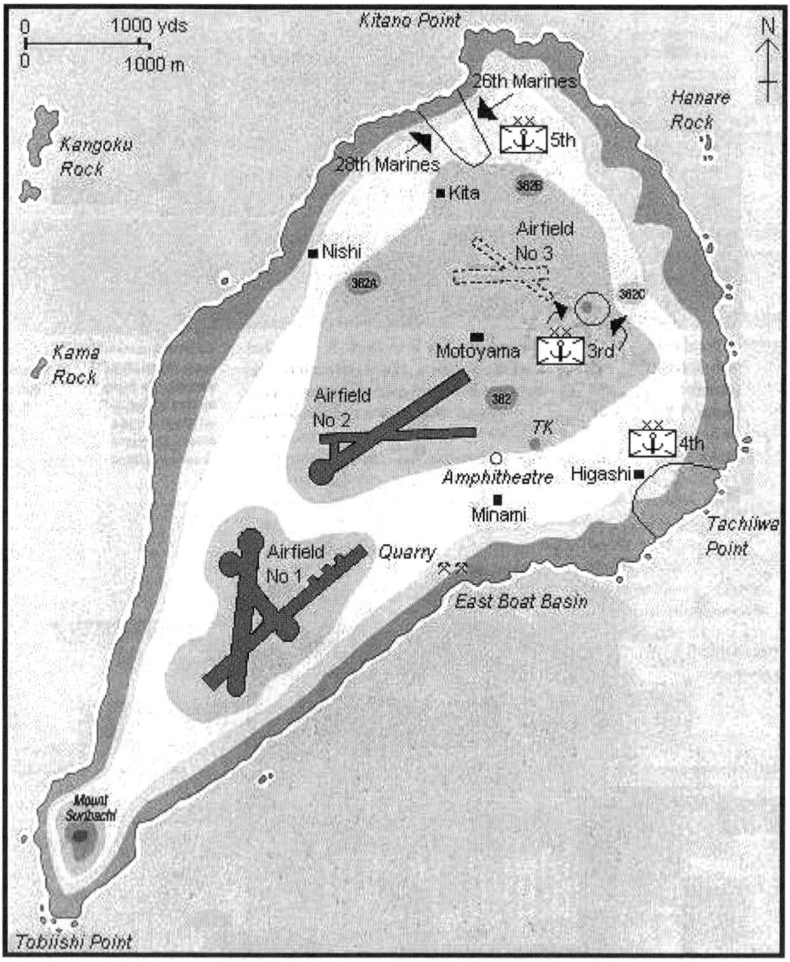

Map 3.1: Gorge Location

(Source: "Battle of Iwo Jima Map 8: D-Day+26," historyofwar.org, ©.)

Not long thereafter on Hill 382, rifleman Jacobson single-handedly took out 16 pillboxes and one buried tank with his newly acquired bazooka. "Much of the regiment's advance that day was due to Jabcobson."[32] However, many of its sub-units were still forced to pull back for the night. That's because all captured pillboxes had been preregistered for enemy artillery and mortar fire. A similar scenario was to replay—over and over—near all Meatgrinder bastions.

> The same sort of thing was happening at Turkey Knob and the Amphitheater. Every time the Marines drove deep into the enemy defenses, they were made to pay dearly for it. The Japanese simply pulled back and called down the mortar and artillery fire which had been "zeroed-in" on their own positions. After the Americans were forced out at dusk, the Japanese returned to prepare a bloody encore for the following day.[33]
>
> — *Battle of Iwo Jima,* by Leckie

Sub-Surface Enemy Movement Had Not Been Blocked

The combination of pinpoint shelling and hidden passageways had made captured Japanese ground nearly impossible to consolidate. Yet, few U.S. troops were to enter any catacomb while fighting still raged above it. Even at the Gorge, they would focus on shrinking the above-ground cordon. In the process, they destroyed a hilltop blockhouse, poured fuel down some holes, and then sealed off all visible entrances to any underground labyrinth. (See Map 3.1.) But, in so doing, they had little affected the defenders' alternate way of moving around.

Only, when the Marines had departed the island and 147th Infantry taken over its consolidation did some of those big underground spaces get searched, and then only cursorily. But, the full extent of the first underground level was never discovered, nor was there any attempt to find subsequent tiers. That first series of rooms and passageways undoubtedly had trapdoors leading to stairwells, as in Figure 2.5. So, the Battle for Iwo eventually hinged on what its liberators considered to be a non-important third dimension.

A Peripheral But Still Important Issue

At Iwo Jima, the Japanese had conducted a 3GW-oriented defense that looked to Westerners like overly lucky 2GW. Their indirect fire had been so well preregistered as to cause many U.S. casualties. But all artillery and mortar rounds had been carefully expended, and often accompanied by unconventional ground operations. Much of the Nipponese maneuvering had occurred piecemeal and below ground. It had channeled limited assets to where the Americans were actively attacking and then shifted them backwards. After repeatedly countering the invader's assaults with 360° ambushes, the defenders had gone after his command and logistics networks with tiny infiltration teams. Then, through what had appeared to be expert artillery work, many U.S. commanders had been killed, and resupply facilities destroyed. All the while, the Japanese had been dodging most of TF 56's massive firepower through camouflage, dummy positions, heavy construction, and reverse-slope positions.

The Japanese had feigned 2GW to preserve the traditional U.S. faith in it. Troops who are worried about a frontal *banzai* attack will be less likely to notice unauthorized personnel to their rear. The whole well-planned progression had been sadly ingenious, as U.S. casualty statistics now bear out.

But the story of Iwo Jima does not have to end there. Combat will always involve a thinking foe. Modern-day U.S. infantrymen and commandos may want to try some of these same kinds of things next time they find themselves badly outgunned. Why then, are they not regularly briefed on the 3GW capabilities of former adversaries? Granted, a steady diet of high-tech 2GW would better protect headquarters control and arms-industry profits. But, when one's life is at stake, those are no longer priority issues. And 2GW, of whatever degree of technological sophistication, has limitations. Its singular purpose is still to kill enemy soldiers. Such a strategy remains no longer comprehensive enough to decisively win an extended ground conflict. That's the ongoing lesson of every big overseas U.S. deployment since WWII.

Why Western military bureaucracies try so hard to protect the initial version of an enemy engagement is not quite clear. Additional information continues to surface on most. In the case of Iwo Jima, no further insight could in any way lessen the accomplishment of its participants. They had been repeatedly knocked down, and then gotten back up to win the bout. In the parallel worlds of pugilism and war, there is no more prideful occasion.

God and truth go hand in hand. Over 80,000 fairly religious Americans would end up serving on God had helped them to win the battle, and they would want all of its particulars fully aired—if only to warn grandsons and granddaughters of similar circumstances in the modern world. To properly honor them and their Creator, this book will cover officially acknowledged details of the battle in a slightly different order—in search of frontline U.S. innovations. Two of the world's best military outfits had fought. Contemporary American gladiators deserve to know what their low-ranking grandfathers had done to win.

4 How Far Had Corps Small-Unit Maneuvers Evolved

- Had infantry assaults been made less risky during WWI?
- How had noisy surprise been harnessed?

Former Raiders had been dispersed throughout the Fleet.

(Source: Drawing by Donald L. Dickson, in "First Offensive," by Henry I. Shaw, Jr., Marines in WWII Commemorative Series pamphlet by HQMC,1992. p. 25.)

USMC Had Planned to Take Iwo with Attrition Methods

By 1945, the Eastern armies of Europe had already replaced—with tactical surprise—much of the bombardment that had been formerly used on enemy fortifications.[1] But the Corps still believed firepower-oriented 2GW to be its only option on tiny Iwo Jima. There, foe pillboxes/caves and their protective machinegun nests would be so closely arrayed that no strongpoint or sub-element could be assaulted without incurring the fire of others. And whenever the Marines happened to bypass a pillbox cluster, they would run the risk of being cut off and annihilated by greater numbers.

But, as the battle progressed, that steady application of 2GW soon included—ever so briefly—impromptu occasions of 3GW. Without them, the final casualty count would have been much higher.

Occasional Shifts in the Style of Warfare

As the 5th Marine Division fought its way north along Nishi Ridge and the western side of the Motoyama Plateau, it encountered terrain so rugged and well-fortified that a slow and methodical advance seemed its only option.

> The infantry advanced until pinned down by fire. Then, when hostile positions were located, armored dozers, working ahead of the line under protection of tank and infantry fire, prepared a road over which the Shermans could advance to engage the enemy. After flame tanks had neutralized the area, the infantrymen advanced while demolition teams dealt with bypassed caves and pillboxes. This was a slow and laborious process, but any [conventional] attempt by the infantry to continue the advance before destruction of enemy positions resulted in excess casualties.[2]
>
> — *Iwo Jima: Amphibious Epic*
> Historical Branch, HQMC

But, not all frontline commanders thought such a time-and-ammo-consuming procedure to be always necessary. They longed for more momentum. That's when Gen. Erskine of the 3rd Division started to look a little harder for alternatives. One of the best was being daily demonstrated by frontline troops with enough courage to try it.

The pulverizing effect of U.S. bombardment had not yet exposed enough of Iwo's fairly level terrain to those who defended it. A blanket of fragmented rock and dead vegetation still concealed many of its ripples. This gave Western attackers the opportunity to operate at a smaller scale than was customary. Down in the watershed crevices and blasted litter lay pathways, up which tiny U.S. maneuver elements could sneak. And hundreds of such forays by one or two volunteers at a time would ultimately prove the salvation of the hard-pressed divisions.

Individual Training Up to This Point in the War

When the 3rd Marine Division hit the beaches on Iwo Jima, "every intricate detail of wiping out pillboxes, concrete emplacements and heavily fortified positions had been [already] studied and rehearsed."[3]

At least part of that troop training had involved crawling (roughly on line) along flat ground and through token barbed wire and explosions. Overhead automatic weapons fire may have been occasionally added to the demonstration. But the machinegunners of some rifle companies had been specifically shown how to shoot just over the right shoulder of anyone rushing a bunker. So, an upright zigzagging approach may have been standard operating procedure for many outfits.[4]

The New Reality and Its Strategic Implications

The U.S. troops who first hit Iwo had been scared. They had heard enough "scuttlebutt" to know that the chances of a walkover were slim, despite the massive bombardment. To make matters worse, all had just been smeared with white anti-flash cream. So, while coming in by Higgins Boat or amtrac, the slightly seasick Marines looked to each other like ghosts. Not long after landing, they collectively concluded that they were in for the fight of their lives, and that winning would take their own hand-to-hand combat with any number of Japanese soldiers. The problem then became getting that close to one. Not only were Marine units pinned down much of the time, but defenders had started to vanish before any assault line could reach them.

So, the rank and fire—under the insightful guidance of experienced enlisted leaders—became determined to personally close with the opposition the first chance they got. Then, it would be a desperate fight to the death with whomever they encountered. Any other course of action (like a stand-up unit assault) would be equally dangerous, so they decided to take personal responsibility for winning the overall battle. The nonrates' motivation would not be enemy hatred or personal satisfaction, but rather a sacred duty to relatives, country, and each other (*agape* in Christian terms). (See Figure 4.1.)

Figure 4.1: Many Marines on Iwo Had Been Religious
(Source: Courtesy of Edward Molina, ©.)

Tiny-Element Stratagems Start to Appear

When the Marines had first landed, their organizational doctrine called for the "blind 'em, burn 'em, and blast 'em" technique for all foe bunkers (with the order of the last two steps sometimes varying). To a rear-area planner, such a sequence made perfect sense—block the defenders' line of sight, neutralize them with flame, and then destroy their emplacement with explosives. But, while the sequence itself was logical, how it was executed soon became an issue. The version applied to a Tawara blockhouse in the movie "Sands of Iwo Jima" would take too long to organize and then been exceedingly dangerous. Though a U.S. machinegunner had been firing into that blockhouse's aperture, the first fuel carrier got shot, and then the first two explosive toters.[5] On Iwo Jima, the enemy bunker would have to be more obscurely reached through a less complicated progression. Soon after arrival at the new beachhead, there were any number of Marines eager to try their rendition. It would often involve fewer participants, less visible movement, and not as much ordnance.

> 19 February 1945. Landing in the first wave under intense hostile machine-gun and 47mm. gun fire from a pillbox to his direct front, Private Melville [of C/1/27] . . . voluntarily proceeded forward of the front lines toward the Japanese position and, gallantly exposing himself to the rain of enemy fire, advanced close enough to the pillbox to use grenades.[6]
>
> — Navy Cross citation for Pvt. Max E. Melville

As the battle progressed, the exact route and type of forward motion became key factors in every bunker assault. The most logical avenue of advance was at the very edge of the attacking unit's assigned lane (while a machinegun at the lane's center held the quarry's attention).

> February 21, 1945. . . . [W]hen his platoon was held up by a sudden barrage of hostile grenades while advancing toward the high ground northeast of Airfield Number One, Sergeant Gray [of A/1/25] . . . quickly moved forward alone to reconnoiter and discovered a heavily mined area extending along the front of a strong network of emplacements joined by covered communication trenches. Although assailed by . . . gunfire, he cleared a path . . . through the mine field, . . . then . . . volunteered to initiate an attack while being covered by three fellow Marines. Alone and unarmed but carrying a twenty-four pound satchel charge, he *crept up* [on] the Japanese emplacement, boldly hurled the short-fused explosive and sealed the entrance. Instantly taken under machine-gun fire from a second entrance to the same position, he . . . braved the . . . fusillades to *crawl* back for another charge, returned to his objective and blasted the second opening. . . . Repeatedly covering the ground between the . . . enemy fortifications and his platoon area, he systematically approached, attacked and withdrew under blanketing fire to destroy a total of six Japanese positions. *(Italics added.)* [7]
>
> — MoH citation for Sgt. Ross F. Gray

Safely to approach an enemy emplacement, one or two people had to crawl along an indirect route through the microterrain, as opposed to conducting a direct stand-up rush. Then, a satchel charge would almost guarantee a kill. (See Figure 4.2.) In other words, the flame step became optional. The overly deliberate "blind 'em, burn 'em, and blast 'em" procedure had been too easily observed by both quarry and supporting guns.

> 21 February 1945. Determined to knock out a powerful and stubborn Japanese gun battery, Sergeant Queeney [of 1/28] dropped to the ground and, crawling slowly, made his torturous way through the shifting volcanic ash under the relentless slash of bullets from supporting enemy pillboxes

Figure 4.2: Buddy Team After Depositing Satchel Charge
(Source: Charles Waterhouse painting, as a part of "Closing In," by Col. Alexander, Marines in WWII Commemorative Series, HQMC pamphlet.)

delivering covering fire for the huge coastal installation. Immediately upon gaining his objective, he thrust a deadly demolition charge into the entrance of the position, shattering the enemy strongpoint.[8]

— Navy Cross citation for Sgt. Martin J. Queeney

Parent Unit's Initial Approach Had Also Been Dangerous

The enemy had no "defense lines" *per se* on Iwo, just successive and slightly offset rows of strongpoints. Their respective machinegun nests could protect all forward positions with crisscrossing fire. Just to get close enough to contest such an array, everyone in the lead platoons of a U.S. assault company had to use all available cover. Then, to escape the bullets from bypassed spider holes, they needed a rearward contingent.

The northeasterly heading divisions tried to keep all forward battalions side by side—to prevent gaps for the enemy to exploit. After dark, each battalion then pulled in most (if not all) outposts to permit maximum firepower. When the word came to move out the next morning, many battalions deployed two companies up and one back, and their companies assumed the same formation. Such were the realities of moving forward on Iwo Jima.

Same Thought Process Applied to the Front Lines

Most lead-echelon squads of an attacking unit would try to suppress all enemy fire coming down their own lanes, whether it had been directed at them or not. That was so sister squads could close with bunkers that such fire protected. But, then something out of the ordinary happened. The latter group started outflanking their quarries. They did so at the very edge of their assigned lane to keep from taking friendly fire from the back. And, whoever had gone forward skipped the first two steps of the bunker assault method to simply toss in grenades. Soon, this "procedural modification" had become more the norm than the exception.

Official film footage of the battle would later reveal one or two young Marines working far forward of their advancing unit. They would then grenade all enemy holes—from the side or back—that they encountered on the way back via a different route.[9] Many of the enemy machineguns that had held up the advance thereby ceased to exist, and the unit could again move forward. Rifleman logic and frontline "trial and error" had just stumbled upon a important 3GW stratagem. It was to accomplish the normal task of an entire unit with only a few people. The savings in lives would be extraordinary.

Smoke grenades were not widely used in this initial approach to an enemy emplacement, because they would have also blinded the friendly base of fire. (On Hill 382, smoke would facilitate only a pullback.[10]) Whenever a particularly tough enemy fortification was encountered, a flame thrower, bazooka, or pole charge soon appeared on the scene. But, its carrier would also have to follow every fold in the ground. The frontline fighters had been demonstrating the proper scale in which to operate against an accomplished defender.

The Increasing Value of Microterrain

The proliferation of crevices and mid-sized rocks on Iwo's central plateau had created automatic cover for all Marines permitted to crawl through it. Few historians have paid much attention to such mundane matters, but career infantrymen know them to be important. Even, the so-called "patrol" up Suribachi had been an Indian file assault along the path of least resistance. It had been made possible by four things: (1) operating in column; (2) seldom standing erect; (3) using runoff ditches or rock outcroppings for cover; and (4) quickly returning any fire.

> Those [Japanese defenders] who remained popped out of holes and caves to resist Schrier's advance only to be cut down. The Marines worked warily up the steep northern slope, sometimes resorting to crawling on hands and knees.[11]
>
> — *Closing In: Marines in the Seizure of Iwo Jima*
> History and Museums Division, HQMC

A Partial End to the Multi-Unit Sweeps

The extent of enemy fire near one of those Japanese defense

belts had been so intense that fully deployed U.S. units couldn't get anywhere near one during the day without excess casualties. Just to get close enough to attack its forward-most emplacements, they were forced to conduct a piecemeal advance of tiny composite elements. Eventually, small penetrations would be possible. A religious U.S. platoon leader would show everyone how the new game was played at the base of Suribachi.

> 21 February 1945. When ordered to attack across open terrain and dislodge the enemy from a series of strongly-defended pillboxes and blockhouses at the base of Mount Suribachi, First Lieutenant Wells [of E/2/28] placed himself in the forefront of his platoon and, leading his men forward in the face of intense hostile machine-gun, mortar and rifle fire, continuously moved from one flank to the other to lead assault groups one by one in their attacks on Japanese emplacements. Although severely wounded while directing his demolition squad in an assault on a formidable enemy blockhouse . . . , he continued to lead his men until the blockhouse was destroyed.[12]
>
> — Navy cross citation for 1stLt. John Keith Wells

The Marine Corps would later change the wording of Wells' accomplishment. He had led "demolition teams" from one bunker to another, finally accounting for 25 emplacements.[13] And his "assault groups" had been reinforced infantry squads. Each contained one or two people trained in demolitions (possibly engineers). Its other attachments may have been a flamethrower man or bazooka man, because that's who another company commander needed for his assault group demonstration.

> 21 February 1945. . . . Capt. McCarthy [of G/2/24] acted on his own initiative when his company advance was held up by uninterrupted Japanese . . . fire . . . . Quickly organizing a demolitions and flamethrower team to accompany his picked rifle squad, he fearlessly . . . charged a heavily fortified pillbox . . . personally hurling hand grenades into the emplacement.[14]
>
> — MoH citation for former 1stSgt and
> rifle company commander on Saipan
> Capt. Joseph J. McCarthy

Again evident is the size of unit assigned to a bunker take-down—the reinforced rifle squad. Capt. McCarthy had simply lent a little moral support to his first squad to try it. As with the assaults by individuals, grenades had played a key role. SNCOs would soon demonstrate how each squad member should behave on such a mission. Most had been veterans of other island campaigns.

> 21 February 1945. . . .Platoon Sergeant Balducci of [F/2/28] reorganized his unit and initiated a determined attack against the hostile forces at the base of Mount Suribachi. Courageously exposing himself to enemy fire to determine the location of hidden machine-gun and mortar nests holding up his company's front, he personally destroyed two machine-gun nests with hand grenades and contributed to the destruction of five mortar pits and connecting trenches before he fell, mortally wounded.[15]
>
> — Navy Cross citation for P/Sgt. Paul B. Balducci

The annals of history attribute no more significance to the "close enemy contact" of such individuals in charge than leadership by example or acts of heroism. But, their unselfish acts may have—at times—been much more than that. They may have been long-overdue modifications to squad procedures or delegations of sufficient authority to squad members. Buddy teams did, after all, make a smaller target than a whole squad or platoon. Wells had been a "spiritual man" who had attended Private Boot Camp at Parris Island before going to Officer's Candidate School in Quantico."[16] McCarthy had known how vital one man could be to firefighting. And Balducci had come from the south side of Chicago, where one's closest pals can make all the difference.

The Official Contribution of Reinforced Infantry Squads

The overall push up the island soon became a type of grinding motion in which no one part of the formation got too far ahead of another. All along that slowly moving front were hundreds of rifle squads that had been individually trained—with their attachments—for bunker take-downs. The extent of their combined effort became more apparent as the key northern bastions were encountered.

The bloody assault to the crest of rough and craggy 362-B was launched at 1600. The two companies burned and blasted their way to the top, using rocket launchers, flame-throwers, and demolitions to the fullest extent to overcome the dense network of interconnecting caves pillboxes along the southern and western slopes. . . .

[Later,] . . . assault squads were still blowing cave entrances and using flame throwers on remaining points of resistance.[17]

— *Iwo Jima: Amphibious Epic*
Historical Branch, HQMC

Company G [2/24] battled throughout the day to dislodge the tenacious enemy from his shattered positions on Hill 382. Flamethrowers, bazookas, and grenades were used in close combat, and assault squads exposed themselves to intense and accurate small-arms fire as they blasted cave entrances in a systematic yard-by-yard advance.[18]

— *Iwo Jima: Amphibious Epic*
Historical Branch, HQMC

Many Large-Unit Attacks Were Now in Column Formation

Situation maps of the battle show division sectors being subdivided into battalion lanes. (Look back at Map 2.1.) But just as a platoon had circumvented the pillboxes on the way up Suribachi, those Marine battalions were soon outflanking strongpoint clusters in the plateau region. As on Suribachi,[19] the preferred formation for such "maneuverability" was a column. Columns could more easily change direction, and had rearward elements to handle bypassed opposition.

On the left of the 4th Division zone of action, the 23rd Marines jumped off at 0800 in a renewal of its assault against Hill 382. Where he had attacked the day before in a column of battalions, Colonel Wensinger now split his zone and committed the 1st Battalion on the left of the 3rd, both units in a column of companies.[20]

— *Iwo Jima: Amphibious Epic*
Historical Branch, HQMC

Now that outflanking strongpoint clusters had become the best way to limit friendly casualties, one battalion sometimes attacked at right angles to, and across the front of, another.[21] That way, it had a ready made base of fire that would automatically stop shooting whenever a sideways-moving Marine appeared in the impact zone. But within any company-sized attack, tiny lead elements still played a crucial role. That's because a buddy team could hide in rubble that would not obscure anything bigger. Wherever Japanese machinegunners were present, the size of the target mattered.

> For the remainder of the day Company C [1/24] inched forward [in tiny increments] through crevices and gullies, blowing up caves and working under constant small-arms fire.[22]
>
> — *Iwo Jima: Amphibious Epic*
> Historical Branch, HQMC

The One-Man Assault Has Its Advantages

Because 19-year-olds are generally more agile than veterans, many of the new guys were soon wanting to try a lone-wolf adventure of their own. That such a thing would ultimately risk fewer lives was why commanders allowed it. The enemy shelling had been so intense that whole U.S. squads were getting blown to bits.

> Early in the morning of the 20th [of February], Capt. McDaniel [of C/1/23] wanted to reposition "Red" Manning's mortar squad and sent word for him to get ready to move out. I had a foxhole . . . with Red's mortar squad . . . and decided to go with them. We were walking single file diagonally up the slope—[all] six of us, Red in the lead and myself in the rear. . . . We started to traverse the narrow lip between two bomb craters when we "felt" artillery shells coming. Instinctively, all six of us dove into the craters. Our actual positions made me dive into the left while the five men from the mortar squad dove into the crater on the right. A tremendous explosion blew sand over me. As I was shaking off the sand, a chunk of "Red" Manning's arm landed next to me. . . . That was the biggest piece left the five men. Sgt. Kraft, a few minutes later, came over to the far side of the

> crater and shouted to me, "Was Manning in that hole?" I answered, "Yes, and his whole squad!" [23]
>
> — C/1/23 public-domain website

This new way of moving forward made sense to the company-grade officers who observed it firsthand. Why not fully utilize the world's best fighting man?

Hand-Placed Charges Were the Key to Pillbox Destruction

The most important phase of any bunker assault was getting an explosive through its gun slit. Grenades were in greatest supply, so they were first tried. But, satchel charges (and sometimes a trained engineer) were almost always needed for a heavy structure.

> [O]n 20 February 1945. . . . Leading his squad through intense automatic fire from the concealed enemy in a relentless assault on three mutually supporting Japanese pillboxes, Sergeant Crow [Demolition Squad Leader in G/3/28] succeeded in reaching his objective despite two casualties and, skillfully maneuvering his men to destroy two of the hostile emplacements, annihilated the third before he fell, mortally wounded.[24]
>
> — Silver Star citation for Sgt. Joe. R. Crow

There Had Been an After-Hours Raid Early in the Battle

It wasn't long before attacking Marines were wishing for less daylight.

> [A] Demolition Corporal . . . of Company I, Third Battalion, Twenty-fifth Marines . . . [was] in action against enemy Japanese forces on Iwo Jima . . . [on] 19 February 1945. Realizing the necessity of demolishing two enemy cave positions which were holding up his platoon's advance, Corporal Sivicky voluntarily entered enemy territory under cover of darkness to place demolition charges in position and, although seriously wounded by a hand grenade while ap-

> proaching the emplacement, succeeded in destroying both strong points and enemy personnel therein.[25]
>
> — Silver Star citation for Cpl. Edward J. Sivicky

Units Had Avoided Trouble to Cross the Isthmus Quickly

Full-strength U.S. battalions had encountered a buzz-saw on the very first day at Iwo. Just to survive, they were forced to conduct an impromptu version of 3GW. They had started out trying to neutralize every pillbox they encountered. Then, when their direct fire had little effect, they outflanked the position with a few people. But, to follow this procedure for every pillbox or machinegun nest was time consuming. So, the battalions trying on day one to sever the island and take Airfield No. 1 left many emplacements to mop-up teams.

> . . . [T]he leading companies had their orders: get across the island fast. Accordingly, E and F Companies [2/27] by-passed as many Japanese installations as possible, knocking out only those that threatened to hold up their advance. Flamethrower teams and grenade-tossing riflemen killed the Japanese inside while engineer teams shattered the emplacements with explosive charges. . . .
>
> Behind the assault companies of the two battalions, reserve companies cleared the by-passed pillboxes and dug out Japanese machinegun nests.[26]
>
> —*WWII History of 5th Marine Division,* by Connor

Dodging Foe Fire Had Led to Better Individual Movement

This original dash had provided two important lessons. The first was to bypass enemy resistance wherever possible. The second was for each rifleman to defeat the first sniper bullet through some individualized "juke." Crossing the isthmus in a hurry had thus come up with some non-standardized on-the-job instruction that was soon to pay huge dividends. Atop Iwo's northern plateau, the terrain was so broken that only fireteam maneuvers would be possible. That's when U.S. riflemen might again need some fancy movement techniques.

> They crept and crawled, dodged and ducked, slithered and staggered—but they moved forward.[27]
>
> — Bill Rose, a Marine combat correspondent

After such a freewheeling display of offense on the first day at Iwo, the SNCO "chain of command" may have seen their chance to help. From that point forward on any approach march, all lower ranks would move a little quicker, duck a little more, and "hit the deck" before any enemy could acquire a three-second sight picture. Movie footage does exist of a small Marine column zigzagging—at a trot—between tiny promontories in what appears to be an approach march. But, most 2GW assault procedures do not allow for any participant to get all the way down. Yet, Marines had crawled during the patrol-like assault up Suribachi, and sooner or later they would want to try it again on line.

Eluding Enemy Machinegun Fire

After a while, other problems began to surface within standard U.S. assault procedure. The natural inclination to avoid automatic-weapons fire had become more pronounced on Iwo. Why face interlocking streams of hot lead when not absolutely necessary? Getting past a cluster of gun nests took only a long-range infiltration like what Schrier had run up Mount Suribachi. A similar patrol had made it to the top of Hill 362-A; and another to the crest of Hill 382 before too many bypassed pillboxes were spotted.[28] But, those three daring reconnoiters had still helped parent units to finally secure attack objectives at a minimal cost in lives. So, initially taking the path of least resistance to discover objective makeup had proven quite productive.

The only difference between long-range and short-range infiltration is their scale. While one wends its way through unit outposts, the other sneaks between individual fighting holes. And the latter has long symbolized the UW portion of a 3GW offense.

Bottom-up Pressure for Less Costly Maneuvers

After U.S. companies had lost many of their commissioned and non-commissioned officers, their remaining leaders became largely surrounded by nonrates. The acting CO of at least one company may have—himself—been a PFC, because there were many E-2 platoon leaders around.[29] There is also evidence of up to 2000 2nd Lieutenants coming ashore as replacements.[30] But, for whatever reason, the average U.S. company commander was now under a lot of pressure from low-ranking subordinates to use safer maneuvers. His parent-unit superiors were still mostly committed to established procedure. But, depending on their knowledge of frontline tricks so far in the war, some were about to follow some grassroots advice. That advice was to forgo some of the mistake-prone, defender-alerting, and largely ineffective bombardment, in favor of more surprise-oriented small-unit maneuvers.

At this time in the Corps, there had been several subtle movements afoot to improve upon "how enemy fortresses were conventionally assaulted." Marine Raiders hadn't needed any airpower or tanks to get three-fourths of the way through the prototype of Iwo's strongpoint belts at Bairoko, New Georgia.[31] Units (like that of John Basilone) had interchanged the second and third steps of the "blind 'em, blast 'em, and burn 'em" technique.[32] And hand-held demolitions had largely taken the place of artillery rounds that had been hardly scratching the walls of Kuribayashi's reinforced concrete structures.

Was Frontline Logic the Only Source of Such Insights?

The CO of 28th Marines—Col. "Harry the Horse" Liversedge—had commanded 1st Marine Raider Regiment at Bairoko.[33] Then, on the first day at Iwo, his units raced across the narrowest part of the island to sever its garrison. In so doing, they had circumvented many pillboxes, and the casualties that frontally attacking them would have produced. Such a push had been much more daring than what most U.S. regiments would have tried. And its example would help to shape a winning strategy for Suribachi and many of the northern bastions.

Sgt. Schrier of E/2/28 had been an enlisted member of Carlson's (Chinese-Communist-modeled) 2nd Raider Battalion on Guadalcanal, so he knew how much damage a single fireteam could do to a larger unit. And Sgt. Michael Strank (who brought Schrier the second flag atop Suribachi) had been a member of 3rd Raider Battalion at Pavuvu (with Liversedge as CO).[34] That makes both of them well versed in light-infantry matters.

The fall of two other Iwo bastions can also be traced to either the Maoist-oriented 2nd and 4th Raider Battalions, or the light-infantry-oriented 1st and 3rd. 1/28 had squads filled with former Raiders. It was to help 2/28 finally take Hill 362-A.[35] Even its MoH winner—Tony Stein—had been a Raider on Bougainville.[36] And 1/28 would also be in on the collapse of the Gorge pocket.[37]

So, the combined effect of former Raiders on the final seizure of Iwo Jima would be considerable. Even the war posters of the period understood the Raider contribution to final victory. (See Figure 4.3.) Yet, HQMC has yet to admit that the Maoist Raider battalions had been practicing 3GW at the small-unit level.

What the Opposition Had Attempted

In the true spirit of 3GW, the Japanese had gone after the Americans' logistics and command networks. But, no matter how many U.S. ammunition dumps were blown or leaders injured, the Marine rank and file kept coming with their grenades and new Raider formula of "working together." By simply cooperating with each other at the squad level, they no longer needed anyone barking orders.

Enough Small-Unit Authority Was Finally Delegated

The lesser effect that rank then had on all Marine operations helped with frontline support. There now existed a much larger source of rear-area security elements. Foe defenders who can disappear whenever attackers approach, also have ways of secretly penetrating friendly lines.[38] That, in turn, leads to diminished resupply.

Only with an interconnected blanket of two-man sentry posts (SPs) could the Marines' behind-the-lines resupply, reinforcement, and evacuation activities have been adequately protected. These SPs were the rear-area equivalent to the "numerous ambushes . . . by the 147th Infantry every night [after the Marines had left the island]."[39] The idea had undoubtedly come from the last Marine unit responsible for island consolidation—the small-unit-oriented 9th Regiment.

With Each Morning Came Another "Do or Die" Exercise

For pinned down Marine units again to move forward, tiny elements had to lead the way. The microterrain was their preferred avenue. Such a "do or die" environment produces outstanding tactical technique. More of its specifics on Iwo are what this study hopes to uncover in Part Four.

Figure 4.3: The Leaderless "Gung Ho" Option
(Source: Wikipedia Encyclopedia, s.v. "Flag-Raising on Iwo Jima." from this url: https://commons.wikimedia.org/wiki/File:Seventh-War-Loan-Drive-Poster-1945.jpg.)

Those "Combat Scouts" Already Had Some New Methods

The Marines who dashed across the isthmus while under full observation from Mount Suribachi and the southern plateau didn't do so in an upright manner. They were minimally crouching.[40] And they mostly moved between shell holes, without staying too long in any one of them. They had been trying as hard they could to avoid a predictable action.

> Infantrymen, anxious to push through this maze [of enemy positions] followed a pattern—get up, stumble a few yards ahead, then drop again. Everywhere shell crater conversations were the same: "We're spotted—let's get out of here."[41]
>
> —*WWII History of 5th Marine Division,* by Connor

Indirect-fire observers on Suribachi had already zeroed in all the big shell holes. That's why Marine units went after enemy mortar sites every chance they got. To do so, they combined organic firepower with impromptu maneuver. The NCO in this next citation will later help Tony Stein.

> 19 February 1945. Assuming command of an assault platoon, Sergeant Savage [of A/1/28] ably directed the fire of mortars and 37mm guns to destroy a large enemy mortar position and several pillboxes and blockhouses. Personally executing a frontal attack on two more hostile pillboxes, he destroyed them with hand grenades and a pistol and killed seven of the Japanese, thereby enabling his company to reach its objective.[42]
>
> — Navy Cross citation for Merritt M. Savage

Middle Division Had Been Picked for the Hardest Job

The massive, forwarding Marine formation would eventually have to face the foe's MLR and equally tough Third Defense Belt. 3rd Division would have to penetrate both at their center. It had served with Carlson's Maoist 2nd Raider battalion at Bougainville.[43] That nontraditional experience had provided guidelines for such a difficult task.

> Erskine and his regimental commanders had been constantly on the alert for weak points in the enemy defenses, and when a soft spot was detected the advance was pushed ahead without regard for the progress of adjacent units, with reserves positioned to protect open flanks. Then the breach was widened by launching flank attacks through gaps between assault units and the bypassed enemy left to be mopped up as the division smashed ahead.[44]
>
> — *Iwo Jima: Amphibious Epic*
> Historical Branch, HQMC

Frontline Initiative Had Been Indirectly Encouraged

In 1945, each Marine infantry enlistee had been conditioned to take over the job of his immediate superior, should that person become incapacitated. This tradition had given the lowest ranks license to get more deeply involved in the Iwo chaos. And, as the offspring of Belleau Wood veterans, that's exactly what many had in mind.

> The last training phases also saw extensive command practice given to subordinate unit leaders. . . . Therefore, officers turned over platoons to sergeants, and corporals and PFCs operated as squad leaders.[45]
>
> —*WWII History of 5th Marine Division,* by Connor

All the Tiny Successes Were Finally Noticed by 3rd Division

While attempting to move up the island, the 3rd Marine Division had encountered enemy fire so intense that its armor couldn't advance. But Cpl. Woody Williams knew the answer to such a dilemma. He would go where fighting vehicles couldn't. Adding a flamethrower to his normal load of satchel charges, he then almost single-handedly neutralized one of the most heavily defended bastions in his parent-unit's zone of responsibility.

> [A] Demolition Sergeant serving with the First Battalion, Twenty-First Marines, . . . 23 February 1945 . . . [was] quick to volunteer his services when our tanks were maneuvering

> vainly to open a lane for the infantry through the network of reinforced concrete pillboxes, buried mines and . . . volcanic sands. Corporal Williams daringly went forward alone to attempt the reduction of devastating machine-gun fire from the unyielding positions. Covered only by four riflemen, he fought desperately for four hours under terrific enemy small-arms fire and repeatedly returned to his own lines to prepare demolition charges and obtain serviced flame throwers, struggling back, frequently to the rear of hostile emplacements, to wipe out one position after another. On one occasion he daringly mounted a pillbox to insert the nozzle of his flame thrower through the air vent, kill the occupants and silence the gun; on another he grimly charged enemy riflemen who attempted to stop him with bayonets and destroyed them with a burst of flame from his weapon. His . . . determination and . . . heroism . . . were directly instrumental in neutralizing one of the most fanatically defended Japanese strong points encountered by his regiment and aided in enabling his company to reach its objective.[46]
>
> — MoH citation for Cpl. Hershel W. Williams

Some of TF 56's more senior officers had now started to realize that totally surprising the opposition can be as effective as overwhelming firepower. That's when the momentum shifted—for good—to the most Christian side. Around 7 March, 3/9 would run a very successful pre-dawn attack on Hill 331, thinking it was 362-C. Then, its commander would unexpectedly move on to the correct hill.[47]

But, what made this two-part attack so noteworthy were its details. Though 3/9 was trying to poke a hole in the foe's Third Defense Belt, its commander had—with the commanding general's permission—forgone all preparatory bombardment. Then, with his flamethrower men hard to spot before dawn, his unit had easily swept across 500 yards of fully outposted terrain to the wrong objective. When that same commander should have been worrying about a massive counterattack, he kept going. Despite hordes of now fully awake enemy soldiers on every side, his men doggedly fought their way another 250 yards deeper into enemy territory. The afternoon seizure of the most strategically important high ground in the belt must have been enough to finally demoralize the remaining defenders.[48] From that point forward, there would be no stopping

3rd Marine Division on Iwo Jima. Yet, it would still need "fireteam chemistry" to clear out the compartmentalized jungle of stone that was Cushman's Pocket.

5 DEGREE OF JAPANESE OFFENSIVE ACTIVITY ON IWO

- Did the Japanese launch any banzai charges at Iwo?
- What types of attacks did they conduct?

Most defenders were also skilled at light infantry assault.

(Source: Courtesy of Cassell PLC, from "World Army Uniforms since 1939," © 1975-83 by Blandford Press, Part I, plate 155; DA Pam 550-35 [September, 1991], p. 105.)

The Iwo Jima Protection Plan

The new way of resisting an invasion included many "spoiling attacks." Their specifics may shed more light on how a 3GW defense works. Only known in the West is that it generally entails squad-sized strongpoints that effectively ambush the quarry.

A Rerun of Stalingrad?

A powerful German army had gotten almost to the Volga River

in Stalingrad before Soviet defenders could stop it. Its stalling was not accomplished through the do-or-die charge that one might expect from an authoritarian army. It was by the raids of tiny—loosely controlled and initiative-reliant—elements. Their initial target had been the enemy outpost line. Each night, small packs of Soviet snipers would creep forward to do their damage. In the morning, they withdrew to hidden vantage points. Russian scouts, on other hand, formed killer teams with a more personal approach.

> Parties of three to five scouts used the . . . ravines . . . to infiltrate enemy lines . . . [and] pounce on outposts.[1]
> — *Red Army Resurgent,* by Shaw

The Japanese, with their new style of defense, were to do something similar on Iwo Jima. But, there were not many listening posts (LPs) out in front of Marine lines, because they would have masked the fires necessary to stop a *banzai* charge. So, most Japanese teams may have tried to sneak into Marine lines.

> Some small-scale *banzai* attacks occurred later in the battle, but for the most part the Marines never faced large-scale frontal assaults. Each night, however, small parties of Japanese soldiers, called "wolf packs," conducted intelligence probes, seeking gaps between units, and quietly exacted a toll on Marine outposts. By day, the defenders hunkered down and waited for the Marines to enter their preregistered killing zones, and the enforced discipline made the battle both prolonged and costly.[2]
> — *National Interest,* by Warfare History Network

What Had Taken the Place of *Banzai* Attacks

To conserve manpower, Kuribayashi had disallowed mass attacks. In their absence, most Japanese offensive activity on Iwo could be more properly labeled "infiltration." The Marines had been warned that this more stealthy kind of attack would target tanks, command posts, communications, artillery, and mortars.[3] It could take any of the following forms: (1) small groups moving through unoccupied spaces between units (long-range infiltration); (2) one or two Nipponese combatants sneaking between defenders' holes (short-range infiltration); (3) tiny teams emerging from undiscovered portals in tunnel networks; or (4) *motti*-like harassment with grenades.[4] The latter would be conducted from the edge of a U.S. position.

Yet, Kuribayashi still allowed a few several-hundred-man attacks to maintain the Marine units' focus on the terrain to their front.

The First Night Ashore

Expecting a multi-thousand-man *banzai* attack, U.S. units may have been a little jumpy the first few evenings of the invasion. On the night of 19-20 February, the official intelligence report confirms only two things: (1) a boat load of 25 Japanese coming ashore along the western coast and all being killed; and (2) Regimental Combat Team (RCT) 27 being unsuccessfully counterattacked on its left flank by 800-1000 enemy soldiers around 0335.[5] But, the official battle chronicles have added some interesting details.

> "On the first night at Iwo, there would be no wasteful vainglorious *banzai* attack. Instead, small teams of infiltrators, which Kuribayashi termed 'Prowling Wolves,' probed the lines gathering intelligence [and possibly doing other things]."[6]
>
> – *Closing In: Marines in the Seizure of Iwo Jima*
> History and Museums Division, HQMC

What those prowling wolves were up to deserves a closer look. The next excerpt contains the full truth from a Japanese source.

> There were several enemy attempts at infiltration (28th Marines AR, 17). . . .
>
> Japanese artillery kept pounding the American positions. Shortly after midnight, the enemy scored a direct hit on the command post of 1/23 on Yellow Beach 1. . . . Around 0400, the 25th Marines ammunition and fuel dump went off with a terrifying roar. Two full boatloads of 81mm mortar shells, gasoline, and flamethrower fuel exploded, caving in foxholes for yards around.

> Initially, these disasters were attributed to lucky enemy hits on these vulnerable targets. It remained for a Japanese postwar history to clear up this point. According to the Japanese version:
>
> Instead of all-out desperate *banzai* charges, Kuribayashi organized small packs of prowling wolves—three or four in a pack—which sneaked in at night to enemy depots or concentration[s] of fuel and ammunition and attacked with demolition charges [possibly with timers] and hand grenades. This new tactic . . . proved quite successful at the nights of February 19 and 20. For instance, heaps of 81mm mortar shells of the 4th Marine Division blew up at the southern coast; flamethrower fuel and gasoline at the same coast also burned (Masanori Ito, *Fall of the Imperial Army*, vol. IV, 55-110), in Newcomb. *Iwo Jima,* notes). . . .
>
> . . . All through the bitter night, the Japs rained heavy mortars and rockets and artillery on the entire area between the beach and the airfield. Twice they hit casualty stations on the beach. Many men who had been only wounded were killed (Sherrod, in Newcomb, *Iwo Jima,* 136, 137).[7]
>
> – *History of Marine Corps Operations in WWII*
> Historical Branch, HQMC

Those wolves had been after much needed Marine Corps supplies. But, they had also sought to destroy howitzers and the opposition's command structure.

> Battery E [2/13], the last to make shore, was unable to occupy its position until 0200 the next morning because its DUKWs were under constant pounding by the enemy. One of its 105mm howitzers was destroyed by fire and another slightly damaged during this period. . . .
>
> . . . The commanding officer [of 1/23] . . . and his operations officer were killed by [what appeared to be] artillery rounds falling in . . . shell holes on the beach that served as the CP [command post] . . . . The 3/25 supply dump on the beach was destroyed by direct hits of enemy rockets prior to dawn.[8]
>
> – *Iwo Jima: Amphibious Epic*
> Historical Branch, HQMC

The Night of 20-21 February

One chronicle reports an attack of undetermined strength from the south around 2000 hours being repulsed in the 5th Division zone with the help of Marine artillery. At 0443, the Japanese also counterattacked in the 4th Division zone with an estimated 100 men. Intermittently, 3- and 4-man enemy patrols attempted to infiltrate both zones.[9] Other sources discuss specific locations and actions.

> The enemy was active . . . launching a counterattack with infantry and tanks against RCT 27. At 0443, approximately 100 enemy attempted to infiltrate the lines of BLT 1/25. . . . BLT 2/24 [also] reported a minor infiltration effort.[10]
>
> — *Intelligence Report,* G-2, TF 56, April 1945

> Small enemy groups were active along the beach in front of LT 226 between darkness and midnight. . . . During the night one or two enemy troops infiltrated the lines of the unit [RCT 27] on our right and stabbed several men.[11]
>
> — Unofficial 5th Marine Division website

After-Hours Activity on 21 February

According to the preeminent chronicle, the night of 21-22 February was characterized by harassing indirect fires, local counterattacks, and infiltration in both division zones. RCT 27 was hit on both flanks at 2100 by a counterattack that was stopped about an hour later without any penetration. At 0245, the same unit repulsed an attempted infiltration on its flank. Concurrently, there were two strange radar contacts along the western coast but no boats discovered. 5th Division's lines would remain intact, but a few enemy had either infiltrated or been bypassed making a mop-up operation necessary. In the 4th Division area, about 200 enemy troops formed at Airfield No. 2 at 2330 and began moving towards 3/23. Naval gunfire and artillery were able to smash the group before it neared friendly lines. Along the whole division front, small infiltration attempts continued with negligible results. As had now become the custom, enemy mortar and artillery fire fell on front lines, beaches, and rear areas throughout the night.[12] Within the intelligence report lies more detail.

> The enemy was extremely active during the night of 21/22 February. . . . An estimated 200 enemy at 2300 . . . launched a counterattack against RCT 23 [from the north] which was broken up at 2330. Several minor infiltration attempts were made between 2400 and 0400. . . . RCT 24 described its night "as fairly quiet": 15 rounds of mortar fire were received at 2105, one of them exploding an ammunition dump on Blue 1. . . . RCT 25 received . . . local counterattack on the left front before dawn. The 5thMarDiv also reported notable enemy activity during the night. In the south, the enemy endeavored to infiltrate the right flank of RCT 28 . . . , and at 0530 its right flank was hit by a counterattack. . . . In the north, RCT 27 was hit by counterattacks of an estimated 800 enemy on both flanks at 2045/21, stopping them by 2200 with no penetration. At 2345 a strong infiltration was attempted against RCT 27. At 0335 the same RCT was strongly counterattacked on its left flank by approximately 800 enemy. . . . RCT 26 reported a counter-landing by approximately 200 enemy . . . and a similar landing by 300-400 enemy . . . at 0555. At 0700, the 5thMarDiv line was intact, but undetermined numbers of Japs had infiltrated the lines.[13]
>
> — *Intelligence Report,* G-2, TF 56, April 1945

But the most likely source of a big *banzai* was Suribachi itself. So, the Marines at its base were especially watchful.

> Dug in close to the volcano's steep face, Marines of RCT 28 could hear the defenders moving restlessly above them. Enemy artillery and mortar fire fell in the area throughout the night . . . [but the Marines] were masked from most of the heavy weapons on the mountain. Two night infiltration attempts were repulsed. Men from the 81mm mortar platoon killed some 60 Japanese in front of 2/28 during one of these efforts, and Company C accounted for 28 who tried to move north along the western beach.[14]
>
> — *Iwo Jima: Amphibious Epic*
> Historical Branch, HQMC

> The enemy confined himself to two attempts at infiltrating the American lines in the 28th Marines sector. Men of the

> regiment's 81mm mortar platoon killed some 60 Japanese in front of 2/28 during one of these efforts. Company C accounted for 28 more who, in accordance with Colonel Atsuchi's final orders, attempted to infiltrate north along the western beaches.[15]
>
> — *History of Marine Corps Operations in WWII*
> Historical Branch, HQMC

The foe's waterborne capability along those western beaches had been duly noted.

> [A]t about 2200, uncoordinated counterattacks on the left front [of RCT 26], and to lesser extent in the center, were repulsed by artillery and infantry fire. . . . [A]n attempted counter-landing . . . by . . . 50 men and another . . . by an estimated 400 men during the night [for RCT 27]. . . . At 2335, three enemy swimmers were killed attempting to come ashore on the west beach [of RCT 28]. An attempted dawn infiltration was made against LT-228 and LT-328 but was unsuccessful.[16]
>
> — Unofficial 5th Marine Division website

As perfect cover for all that infiltration activity, the night of 21-22 February had been full of big enemy shells. When Beach Blue 1 came under mortar fire, only 15 rounds had been enough to detonate a 24th Marines' ammo dump. That explosion was severe enough to endanger everyone on Blue and Yellow Beaches. It was also heard all over the island. As had happened at Guadalcanal, many of those "lucky mortar hits" had actually been satchel charges.[17]

After the Sun Went Down on 22 February

According to the intelligence report, there was lighter enemy activity during the night of 22-23 February. In the 4th Division zone, RCT 25 reported "enemy movement along its front and some mortar fire shortly after dawn." At RCT 24, infiltrations involving 200 enemy were attempted. In the 5th Division zone, the enemy tried infiltrations along its entire front, and enemy swimmers reached the western shore to an extent that extensive mopping up became necessary. RCT 28, on the Suribachi front, had a series of attempted penetrations.[18]

The Night of 23 February

A Marine patrol had just gotten to the top of Suribachi during the previous day, but there was still plenty of enemy activity at its base after dark. Some of its remaining defenders tried to break out to the northeast.

> Enemy activity during the night, was limited to small groups attempting infiltration of our lines. BLT 28 reported that Japs from Suribachi infiltrating their zone were unarmed except for hand grenades and demolition charges. It was believed that they were headed for the Corps artillery positions. . . . The CP of the 5thMarDiv was shelled at various times during the night by artillery. . . . Occasionally knee mortar fire was also received from Japs hidden near the CP area, in clusters of our own dead.[19]
>
> — *Intelligence Report,* G-2, TF 56, April 1945

> For the night defense on D plus 4 . . . [much of] the [28th Marine] regiment manned positions around the base of Suribachi. During the hours of darkness, 122 Japanese were killed trying to infiltrate the lines. Most of these had demolitions secured to their bodies and were probably trying to reach command posts and artillery positions before destroying themselves. A few rounds of high-velocity artillery fell in the area.[20]
>
> — *Iwo Jima: Amphibious Epic*
> Historical Branch, HQMC

Some if those Suribachi survivors had been headed for friendly units in the Motoyama defense zones.

> [T]he fight for Mount Suribachi was virtually over. Substantial numbers of the enemy, perhaps 300 in all, still occupied . . . the [inner] volcano. . . . Only half of them decided to remain and fight it out. The remainder crawled out into the

> murky darkness and tried to make their way north through the American lines. Most of them fell victim to accurate fire from alert Marines, determined to halt any infiltration. About 20 . . . made it across the lines and reached General Kuribayashi's headquarters near Motoyama in the northern part of the island where they were reassigned.[21]
>
> – *History of Marine Corps Operations in WWII*
> Historical Branch, HQMC

> Attempted enemy infiltration [of RCT 28] during the night failed.[22]
>
> – Unofficial 5th Marine Division website

Some of the Japanese swimmers who came ashore may have been trying to reach the first enemy defense line just beyond Airfield No. 1.

> During the early morning hours of 23 February a band of Japanese swam ashore on the western beaches and attempted to infiltrate the area of the 27th Marines, in 5th Division reserve. Alert for such action, Marine security elements quickly detected this force but extensive mopping up was [still] necessary. In the 4th Division zone, an estimated 100 Japanese attempted to penetrate the lines of both 2/25 and 3/25 at about 0500.[23]
>
> – *Iwo Jima: Amphibious Epic*
> Historical Branch, HQMC

Summary of Sapper Activity at the Beach

The number of Japanese infiltrators spotted carrying satchel charges would suggest their subsequent involvement in some of the more accurate "beach shellings." The CPs of 1/23 and 5th Division had come under grenade or knee-mortar attack from nearby enemy soldiers. An RCT 24 ammunition dump, 3/25 supply dump, and 25th Marines ammo/fuel dump had all exploded. In close proximity to the last event had also been the loss of two full boatloads of 81mm mortar shells, gasoline, and flamethrower fuel. Additionally, two U.S. artillery pieces had been damaged, as well as the DUKWs carrying others.

As a result of all this suspicious beach activity, the U.S. infantry battalions would have to go without much of their organic fire support (to include 81mm illumination) for the first four days of the battle.[24] The Japanese chronicle reference strongly implies that Kuribayashi's prowling wolves had been directly responsible for this key shortage. And they may have also limited the flamethrower fuel available.

At night, Japanese artillery rounds should have been less precise than in the daytime (harder to adjust onto a specific target from Suribachi). However, suffering a direct hit on Iwo Jima generally took unintentionally occupying a preregistered target. That's because the Japanese artillery chief didn't like the expenditure of extra rounds.

> As a general rule, there was to be no adjustment of artillery fire as practiced by United States artillery units, and emphasis was given to surveys and preregistered fires.[25]
>
> – *Iwo Jima: Amphibious Epic*
> Historical Branch, HQMC

The Possibility of Stormtrooper Technique

The Intelligence Report's collocation of satchel charges with artillery impacts may indicate WWI German Stormtrooper technique as well—secretly penetrating enemy lines by throwing explosives that mimic indirect fire.[26] But, most Marines of this era had never heard of such an assault by a single squad. Only now can its possibility be entertained.

On the night of 24 February, MoH-winning Jacklyn Lucas of 1/26, was awarded a Silver Star for repelling an "infiltration attack" on his company. The number of participants in such an incident would indicate whether or not it was Stormtrooper technique. PFC Antony Muscarella E/2/25 single-handedly killed eight Japanese soldiers attempting to "infiltrate" his company around the same period.[27]

During the U.S. Move to the Northeast

The main attack toward the far end of the island had now

kicked off with two divisions on line. The 5th was on the left and 4th on the right (with 21st Marines temporarily attached). The 5th would be mostly responsible for taking Airfield No. 2. By the 27th of February, all of Airfield No. 2 was in Marine hands, the left end of the line about two-fifths of the way up the island, and the right end still just above the Boat Harbor. Not until 1 March would Hills 362-A and 382 be contested.[28] That's when the enemy began again to strike back.

On the night of 3 March, after taking Nishi Ridge and Hill 362-B, 5th Marine Division troops engaged a sizeable force of infiltrators.[29] But local accounts still mention squad-sized elements going after Marine CPs.

> 3 March, 1945. When his forward position was attacked by twelve marauding Japanese just before dawn, Private First Class Carter [of A/1/28] skillfully shot and killed four of them and . . . while exposed to several more of the Japanese surrounding the installations, killed the [next] attacker in a daring hand-to-hand encounter. . . . Private First Class Carter remained at his wounded companion's side until daybreak, surrounded by the bodies of twelve Japanese, five of whom he personally had killed in his gallant defense of his Company Command Post.[30]
>
> — Navy Cross citation for PFC Andrew J. Carter

By this time, the 3rd Marine Division had entered the fight in the middle of the formation. Before long, its units had advanced slightly farther than those of the 4th. That's when a series of bigger infiltration attempts occurred near Airfield No. 3. Because of the gap between Marine units, they would be of the "long-range" variety.

> Shortly after midnight [on the 3rd], the 1st Battalion, Twenty-first Marines, reported a local Japanese counterattack in its zone of action. Simultaneously, the 2d Battalion, Ninth . . . reported similar activity along its front. . . . A gap of some 150 yards existed between the left company of the 2d Battalion, Ninth, and the right flank of the 1st Battalion, Twenty-first, which was covered only by fire from the two units. . . . At 0130 an estimated two hundred Nips

> began infiltrating through the gap near the eastern end of Airfield No. 3.[31]
>
> —*Third Marine Division,* by Aurthur and Cohlmia

On 8 March, Gen. Senda conducted a breakout of the 4th Division cordon around the Amphitheater and Turkey Knob. This was no small-scale operation.

> [On] the night of 8 March . . . Gen. Senda grew so frustrated by the tightening noose being applied by the Fourth Marine Division that he led 800 of his surviving troops in a ferocious counterattack.[32]
>
> – *Closing In: Marines in the Seizure of Iwo Jima*
> History and Museums Division, HQMC

There was also a massive counterattack by mostly Japanese naval personnel some 500 yards north of Higashi on the night of 8-9 March. (This incident will be further discussed in Chapter 11.) Here, as at Suribachi, there were indicators that spearhead squads had used Stormtrooper technique (satchel charges mimicking indirect fire). 2/23 had been holding the line nearby.

> At 2300 [11:00 P.M. on 8 March] the 2nd Battalion, 23rd Marines . . . reported large-scale infiltration attempts along the regimental boundary in the area of the salient. . . .
>
> . . . A few even infiltrated through the front lines to the 2/23 command post where they harassed the operations section with hand grenades thrown from ranges of 10 to 15 yards. . . .
>
> These enemy troops were well-armed and equipped, and many carried demolition charges. The attack was preceded by an all-out artillery preparation.[33]
>
> – *Iwo Jima: Amphibious Epic*
> Historical Branch, HQMC

Before long, the defenders of Iwo had gone back to their "wolf pack" tactics. On the night of 11 March, 3rd Division received infiltrator activity resembling Finnish *motti* tactics.

> After nightfall the Japanese, employing hit-and-run tactics, emerged from their hideouts and stealthily approaching the

> Marine positions, hurled a few hand grenades, and then attempted to disappear as quietly as they had come. Such tactics succeeded only infrequently, and many of the infiltrators did not survive to tell of their exploits.[34]
>
> — *History of Marine Corps Operations in WWII*
> Historical Branch, HQMC

Around 11 March, there was some telltale activity in the 5th Division's defense zone. "Small groups of Japanese repeatedly attempted to sneak through the Marine lines. A few reached the 81mm mortar positions of 3/27. Seven of the infiltrators were then killed, one wearing a Marine uniform and equipped with an M-1 rifle. Around 2100, an enemy troop concentration across from the 28th Marines was broken up by an indirect-fire barrage that killed 26 of them. Sporadic rifle fire and grenade duels continued for the remainder of the night."[35]

The night of 14-15 March was marked by continuous activity against the 5th Division. Around 0200, close to 100 Japanese attempted to infiltrate the positions of 3/27. About the same time, a small number of Japanese approached the perimeter of 2/26 and started tossing in hand grenades.[36]

The Offensive Nature of Japan's New Resistance Plan

Throughout history, defenses have attempted to disrupt the momentum of an invader. Once slowed down, he is thought to be more susceptible to counterattack.

Yet, the Japanese defensive belts on Iwo were so lethal as to provide their own offense—in the psychological realm. They not only sapped the attackers' combat strength, but also his organizational resolve. To accomplish such a thing in battle, one must normally do three things: (1) totally surprise the opponent; (2) badly hurt him; and then (3) ostensibly suffer no personal injury from the encounter.

Kuribayashi had not always put his defensive positions where a Western adversary would expect them—places allowing for interlocking bands of grazing fire. To counter the threat of tanks and artillery, he often selected partially obstructed or reverse-slope ground that would still have every part covered by some munition. In that way, he could stymie U.S. pushes through the unexpected meshing of defensive fires with broken terrain. The details of how thus to counter a more powerful foe should still be of interest to U.S. infantrymen. Of particular interest is how the Japanese machineguns were employed.

> Enemy weapons and mine fields were well placed to take maximum advantage of terrain. . . . Machine-gun positions controlled [shot down] draws, while mortars were disposed to deliver fire on routes of approach, including the crests, forward, and reverse slopes of ridges.[37]
>
> – *Iwo Jima: Amphibious Epic*
> Historical Branch, HQMC

Then, as in what would later become known as the UW part of 3GW, the Japanese conducted a series of very demoralizing ambushes.

Sub-Surface Forays at the Amphitheater

Not every spoiling attack must be a raid, for it can also lead to an ambush of enemy soldiers. The Japanese on Iwo did more than fight for one place and then withdraw to another. To escape the pre-assault bombardments, they would—at the last minute—move forward through tunnels to secretly occupy defensive positions right in front of advancing U.S. units. Such a stratagem might now be called "offensive hugging." It largely dodged the Americans' preparatory bombardment.

> As a result of a close study of the enemy's recent defensive action, aided by observation from OPs [observation posts] and air reconnaissance, the following explanation is suggested of the enemy's defense in this Div[ision]'s Z[one] of A[ction]. The enemy remains below ground in his maze of communicating tunnels throughout our preliminary arty fires. When the fire ceases, he pushes OPs out of entrances not demolished by our fires. Then choosing a suitable exit he moves as many men and weapons to the surface as he can, depending on the cover and concealment of that area, often as close as 75 yards from our front. As our troops advance toward this point, he delivers all the fire at his disposal, rifle,

machine-gun, and mortar. When he has inflicted sufficient casualties to pin down our advance, he then withdraws through his underground tunnels most of his forces, possibly leaving a few machine gunners and mortars. Meanwhile our Bn CO has coordinated his direct support weapons and delivers a concentration of rockets, mortars, and artillery. Our tanks then push in, supported by infantry. When the hot spot is overrun we find a handful of dead Japs and few if any enemy weapons. While this is happening, the enemy has repeated the process and another sector of our advance is engaged in a vicious firefight, and the cycle continues.

Supporting indications to these deductions are:
(1) When the hot spot is overrun we find far too few dead enemy to have delivered the fire encountered in overrunning the position;
(2) We find few if any enemy weapons in the position overrun but plenty of empty shell cases;
(3) We find many tunnel entrances, some caved in, all appearing deep and well-prepared, some with electric light wires;
(4) During the cycle, close air and OP observation detects no enemy surface movement."[38]

– *4thMarDiv Periodic Report,* 6 March 1945

***Rendezvous*-Like Infiltrations at the Island's Far End**

Near the end of the battle for Iwo, foe soldiers who had individually snuck through U.S. lines would occasionally *rendezvous* in significant numbers for a follow-up attack. During the night of 24 March, 1st Lt. Harold G. Schrier, CO of D/2/28 [and Suribachi patrol leader], rallied his men when a group of about 100 Japanese soldiers "infiltrated through the main defensive positions shortly after midnight." They then launched an assault *en masse* against the rear of his lightly-manned command post.[39]

After the sun had set on 26 March, over 200 Japanese diehards from the Bloody Gorge at Kitano Point conducted one last offensive operation. To escape the Gorge cordon, they are known to have performed a short-range exfiltration. To later reappear inside the Airfield No. 1 perimeter, they must have either snuck in—a few at a time—or used a tunnel. Over the intervening two-and-a-half-miles, they had likely conducted a long-range infiltration above ground. All three feats on the same night would have taken extraordinary light-infantry expertise.[40]

Bare-Handed Armor Killing

Throughout Iwo's 36-day ordeal, Japanese soldiers had many times tried to destroy U.S. tanks at close range. Some rushed out with pole charges, while others tossed smoke grenades before attaching sticky or magnetized explosive charges. There were also satchel charges and Molotov cocktails being thrown.[41] Within WWII Europe, similar methods had already been employed by the frontline infantrymen of both German and Russian armies.[42] Well-documented in the Pacific had also been the pulling of a landmine on a string under the treads of a tank.[43] Without such offensive forays, no GW-oriented defense (series of near ambushes) can succeed against a fully supported adversary.

PART THREE

Organizational Issues

"ALL COURAGE COMES FROM THE HOLY SPIRIT."
— A MOTTO OF THE U.S. NAVY CHAPLAINCY

(Source: Attributed to U.S. Navy Chaplaincy.)

6 THE JAPANESE SOLDIER'S CODE OF CONDUCT

- What were the strengths of a Japanese combatant?
- Did he have any weaknesses?

Japanese code of Bushido played a major role on Iwo.

(Source: Courtesy of Cassell PLC, from "World Army Uniforms since 1939," © 1975, 1980, 1981, 1983 by Blandford Press Ltd.)

The *Senjinkun*

"Instructions for the Battlefield" *(Kyujitai Shinjitai Senjinkun)* was a pocket-sized military pamphlet issued to soldiers of the Imperial Japanese forces on 8 January 1941. It was published in the name of War Minister Hideki Tojo,[1] and in use at the outbreak of hostilities with America.

The *Senjinkun* was regarded as a supplement to the "Imperial Rescript to Soldiers and Sailors" that was already required reading for the Japanese military. This other work listed a number of exhortations regarding the following: (1) military regulations; (2)

combat readiness; (3) *esprit de corps;* (4) filial piety; (5) veneration of Shinto *kami* (spirits, phenomena, or "holy powers"); and (6) Japan's *kokutai* (system of government). The *Senjinkun* code was an attempt to restore the military discipline that had been lost in the Sino-Japanese War of 1937. More specifically, it was to discourage the "atrocious behavior" that had occurred in the Rape of Nanking (also known as the Nanking Massacre). Most of this code was devoted to upholding the Imperial Japanese soldier's honor—avoiding getting drunk, resisting lust, and treating non-combatants with kindness.[2] But, it also forbade retreat or surrender.[3] Its reminder—"never live to experience shame as a prisoner"—was later cited as the cause of numerous suicides by soldiers and civilians, alike. This code also led to higher fatality and lower surrender rates for Japanese ground forces. When faced with defeat, they would often launch suicide attacks or kill themselves outright.[4]

While *Senjinkun* theoretically told Japanese soldiers to "show mercy to those who surrender," its concurrent disdain for personal capitulation may have ultimately lead to the maltreatment of prisoners.[5]

> [I]n the *Senjinkun* . . . soldiers were enjoined to "show kindness to those who surrender."[6]
>
> *War Without Mercy,* by Dower

"Imperial Rescript to Soldiers and Sailors"

The "Rescript" stressed absolute personal loyalty of each military member to the Emperor. It cautioned that member against any involvement with political parties or politics in general, and the influence of newspaper opinions. The "Rescript" also advised all soldiers to be frugal in their personal habits (reflecting back on the *Samurai* tradition), and respectful and benevolent to civilians (mirroring European traditions of chivalry). Yet, a clause that the military was subordinate to civilian authority never made it into the final draft.[7]

The "Rescript" contained a few Confucian themes like "showing proper respect to superiors." It also reflected Buddhist influences: "The soldier and the sailor should make simplicity their [joint] aim."[8]

Code of the Samurai

Bushido had expanded and formalized the earlier Code of the *Samurai,* and stressed sincerity, frugality, loyalty, mastery of martial arts, and honor to the death. (See Figure 6.1.) Under the *Bushido* ideal, a *Samurai* failing to uphold his honor could only regain it through *seppuku* (ritual suicide).[9] Of course, resisting surrender or a foolhardy attack would also work. The *Bushido's* basic principle was "death before dishonor or defeat."

Figure 6.1: Ancient Samurai Warrior
(Source: "Corel Gallery Clipart," Totem Graphics, Man Historical, 28T009)

To counter the enemy's firepower advantage, *Bushido* was further promoted in the Imperial Japanese Army (IJA) during the Russo-Japanese War of 1905. Japan ended up winning that war, but not without heavy infantry casualties.[10]

All Soldiers Follow Through on Their Martial Training

The frontline Japanese soldiers of WWII were heavily oriented towards close-quarters combat. That's because—by doctrine—the Japanese Army only recognized victory to a tactical situation in which its troops had used their bayonets.[11] In fact, to better accomplish surprise during a deliberate night attack, it had specifically disallowed any small-arms shooting.[12]

Westerners of this era commonly believed that all low-ranking Japanese riflemen were either mindless automatons or compassionless animals. Of course, they were neither. But, they were being repeatedly involved in extremely dangerous, small-unit missions. Partially trained in bare-handed tank fighting,[13] they would also take on many times their number of infantrymen. Each squad had been given a semi-autonomous role on both defense and offense. As such, all riflemen enjoyed some degree of initiative.[14] That initiative—in combination with *bushido* and a Sun Tzu heritage—made any Iwo defender quite capable of tossing a hidden grenade after feigning surrender.

7 Japanese Homeland Assistance for Iwo Jima

- How many supplies did foe have on hand for the fight?
- What was the water situation?

Had any Nipponese vessels been reaching Iwo?

(Source: "Corel Gallery Clipart," Totem Ship, 37C019.)

Japanese Survivors of Guadalcanal Had Been Rescued

In late 1942, Japanese submarine I-1 had come to the assistance of the Nipponese solders stranded during the fighting for Guadalcanal.

> During the preparations for the evacuation (which would begin on February 1 [1943]) at Rabaul, the Japanese submarine I-1 loaded up with about 10 tons of Army rations and set out on a relief mission for Guadalcanal on January 24.[1]
> — *National Archives and Records Administration*

The government of Japan would end up removing 10,652 men from Guadalcanal. During that evacuation, 600 died and 3,000 more required extensive hospital care.[2]

One of Japan's Alaska Garrisons Had Also Been Saved

In May 1943, U.S. forces landed on Attu during "Operation Landcrab" and supposedly destroyed the entire Japanese expeditionary force on that 345 square-mile island. On 28 July, the Imperial Japanese Navy evacuated the Kiska invaders, ending all Nipponese presence in the U.S. Aleutian Islands.[3]

> After the heavy casualties suffered at Attu Island, [U.S.] planners were expecting another costly operation. The Japanese tactical planners had, however, realized the isolated island [Kiska] was no longer defensible and planned for an evacuation.
>
> Starting in late July, there were increasing signs of Japanese withdrawal. Aerial photograph analysts noticed that routine activities appeared to greatly diminish and almost no movement could be detected in the harbor. Bomb damage appeared unrepaired and aircrews reported greatly diminished anti-aircraft fire. On July 28, radio signals from Kiska ceased entirely.[4]

Some of the Off-Shore Rocks Had Been Searched at Iwo

Kangoku Rock was a 600-yard-long bit of volcanic debris lying some 2,250 yards off the northwestern shores of Iwo Jima, and Kama Rock lay two thousand yards to its south.[5] The men of Company B, Fleet Marine Force Amphibious Reconnaissance Battalion had searched both Kama and Kangoku on 13 March 1945, but encountered no resistance. Evidence pointed to Japanese soldiers being on the latter several weeks before. There were a few barricaded caves and stone positions, but no equipment. Both little islands sat fairly low in the water.[6]

> D-plus 11. Three batteries of the 506th [Antiaircraft Artillery Gun Battalion, U.S. Army] were assigned the mission

of neutralizing enemy installations on Kama and Kangoku Rocks, which lay off the northwest coast. The 5th [Marine] Division had been reporting rocket and mortar fire from these rocks, and it was planned to silence the positions by naval gunfire, strafing, and bombing. After [a] considerable working over, light antiaircraft fire was still observed, and the Army 90mm gun units were given blanket authority to fire whenever Kama and Kangoku showed any activity.[7]

– *Iwo Jima: Amphibious Epic*
Historical Branch, HQMC

Japanese Submarines Had Been in the Iwo Jima Area

Before the U.S. Navy blockade of Iwo Jima, that tiny island had been partially supplied by submarine. But then, all underwater movement stopped for a while.

19 February 1945: The *"Chihaya" kaiten* [human torpedo] group is formed. The plan calls for I-44, I-368, and I-370 of the "Chihaya" group to attack American ships at anchor off Iwo Jima.

22 February 1945: On the third *kaiten* mission, I-44 departs Otsujima at 0900, on her fourth war patrol following I-368 and I-370.[8]

—Unofficial Imperial Japanese Navy webpage

25 February 1945: 48 nm [nautical miles] SW of Iwo Jima, while recharging her batteries, I-44 is spotted by two or three destroyers or sub-chasers and forced underwater. She is chased for the next 47 hours.

26 February 1945: [I-44] arrives south of Iwo Jima.

28 February 1945: Lt. Kawaguchi [of I-44] attempts to approach Iwo Jima from the east but is attacked by an "Avenger [airplane]." He reports his failure to HQ, Sixth Fleet.[9]

—Unofficial Imperial Japanese Navy webpage

1 March 1945: Lt. Kawaguchi reports that he intends to head for Oki-Daito Shima area.

> 6 March 1945: Vice Admiral Miwa orders all submarine operations in Iwo Jima area canceled. I-44 receives the order to return [home].[10]
>
> —Unofficial Imperial Japanese Navy webpage

Might a Submarine Have Reached Iwo After the Battle?

Once the Marines had left Iwo Jima, most of the blockading ships would have departed as well. Some 50 years later, a rumor would circulate around the E-7 ranks of the U.S. Navy that one of Japan's massive late-war submarines had managed to visit the island at some point.[11] Unfortunately, there is no way of substantiating such a claim. There were 18 I-400 class boats originally planned, but only four ever completed. One of those four was converted to a tanker in June of 1945 and then decommissioned in November. Two others were not captured until August of that year.[12]

The documentary slice of Iwo Jima had intentionally omitted both ends. A submarine pen was not likely at its northeastern end, though volcanic islands tend to have steep shorelines.[13] But, fully possible could have been a water pipe leading out to an underwater ledge some 2000 meters offshore. (See Map 7.1.)

And during the last two years of the war, many Imperial Japanese Navy submarines are known to have transported supplies to island garrisons cut off by U.S. naval and air blockades.[14]

Another Possibility

According to two official U.S. accounts of the battle, the Japanese troops on Iwo Jima had seventy days' supply of rice by February 1945, plus other rations which could be made to last for two more months.[15] Gen. Kuribayashi's farewell telegram had made separate mentions of limited assets: (1) "ammunition gone and water dried up" and "being empty-handed and ill-equipped." Many years later, a Japanese researcher would convert those generalized admissions into the much more specific "no weapons and [all] supplies cut off."[16]

While consolidating the island., U.S. Army units encountered holdouts with plenty of food and water. Their formal report said many Japanese soldiers, whether dead or captured, showed no signs of starvation or privation at all.[17] In way of explanation, all enemy soldiers were credited with being good infiltrators.

So, the enemy's food supply had never really been at issue during the 36-day battle, only his water. And certain locations hadn't run short of the latter until mid-March. The defenders of Senda's pocket had plenty to drink.

> Late in the afternoon on 11 March 1945, a captured Japanese soldier reported that about 300 enemy troops remained in caves and tunnels of the small area of resistance to the front. He told Marine interrogators that this die-hard group had rifles, ammunition, grenades, demolition charges, and plenty of water, but little food. The highlight of this [prisoner of war's] POW's information was that a Japanese major gen-

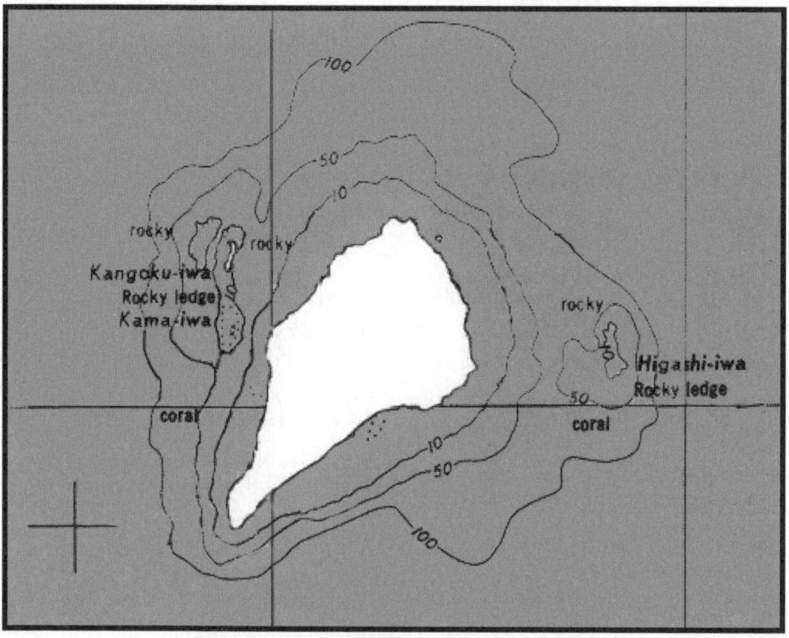

Map 7.1: The Underwater Ledge
(Source: Wikipedia Encyclopedia, s.v. "Volcano Islands.")

> eral held out with the troops in the pocket (4th MarDiv D-2 Periodic Rpt. No. 68, 11Mar45).[18]
>
> — *Iwo Jima: Amphibious Epic*
> Historical Branch, HQMC

One Mystery Still Lingers

During the U.S. Army sweeps of the island, a large hospital was discovered below the Third Defense Belt and not far from the documentary tunnel.[19] The stack of huge chambers at that tunnel's northeastern end would have made a perfect recuperation ward for what must have been thousands of wounded Japanese soldiers. Before long, the American blockade may have sufficiently lifted for enemy submarines to resupply that ward with water (either by pipe or human torpedo). And a secretive, piecemeal evacuation could have ensued.

When the Marines had left the island, over 10,000 Japanese were still unaccounted for.[20] Now that the full extent of Iwo's labyrinth has been revealed, the final entombment of every missing man is far from certain. Many had been in underground defense complexes that had been specifically designed—with more than one level and numerous ventilation shafts—to withstand surface closure.[21] (Look back at Figures 2.5 and 2.11.) Not every shaft had been sealed. To this day, the island is covered with them. So, the trapped Nipponese soldiers had more than enough oxygen to cut their way through the soft rock to a feeder tunnel that would eventually lead to the island-long thoroughfare.

The Japanese headquarters cave in the Gorge had transmitted the following message on 21 March to its rear echelon on Chichi Jima. But, the reference to "not eating and drinking" clearly applies to the Gorge detachment alone. There is no way that all bypassed Japanese soldiers could have just reported a similar plight.

> Then on the 21st Chichi Jima received a message from Iwo, "We have not eaten nor drunk for five days. But our fighting spirit is still running high. We are going to fight bravely till the last." After a silence of almost three days, Hori's radio [on Chichi Jima] crackled again and his operator brought him another message, "All officers and men of Chichi Jima, goodbye." This was the last word from the Japanese defend-

ers on Iwo Jima.[22]

— *Iwo Jima: Amphibious Epic*
Historical Branch, HQMC

Such a dire shortage of life-sustaining supplies could not have been true inside the Gorge or at all bypassed contingents, because the 26 March raid on Airfield No. 1 had yet to be run. And during the months of April and May 1945, the 147th U.S. Infantry would kill or take prisoner some 2500 healthy looking enemy soldiers.[23] So, the Gorge message, on which U.S. intelligence had based its end-of-hostilities assessment, had been an intentional deception. Unless the "still missing" 7,500 fighters had been the figment of someone's imagination, there's more to their story. Wounded foot soldiers with a slim chance of evacuation would have had no obligation to commit suicide.

8 Procedural Challenges for the Marines on Iwo

- Which fire support issues were soon apparent?
- How easy was it to resupply the frontline units?

144 of 168 howitzers ashore couldn't dent a cement bunker.

(Source: U.S. Army Center of Military History, artphoto archives, image designator "0806-3.jpg.")

Effect of the U.S. Bombardment

A noise-and-light show may bolster assault troop morale, but it does little to damage a well-dug-in opponent. There's some doubt as to how much difference seven more days of pre-invasion naval bombardment would have made. If adjusted by aerial observers, it might have penetrated a few of the occupied blockhouses atop the plateau. But, as had happened at the landing beaches,[1] their rubble could still have been reentered from below. And as the assault on Monte Casino in Europe had demonstrated, a rubbled fortress can be just as difficult to take as one still standing.[2]

Toward the far end of Iwo's plateau, the Marines encountered deep draws. Only a U.S. ship sharing their long axis could have reached their bottoms with big shells. Of course, Marine artillery would have the same problem with precipitous terrain, and it didn't pack the punch of a naval gun.

> The 5th Marine Division drive into the biggest enemy pocket remaining on the island [the Gorge] got under way on 11 March. The attack was preceded by a 10-minute preparation, which continued until 20 minutes after the jump-off. The half-hour barrage, fired by the 12th, 13th, and 14th Marines, as well as the corps artillery, provided an impressive spectacle but once again, in accordance with previous experience, was generally ineffective against enemy personnel who huddled in well-protected cave positions.[3]
>
> — *History of Marine Corps Operations in WWII*
> Historical Branch, HQMC

Not only did U.S. indirect-fire weapons have trouble hitting their targets, but they often lacked the power to penetrate walls or roof.

> Standard artillery tactics and doctrine proved sound at Iwo, but the organic [U.S.] weapons were inadequate for the task of destroying the type of emplacements encountered. The 105mm and 75mm howitzers of the divisional artillery battalions were completely unsuited for this work. Even the much heavier 155mm shells of corps units required ten [10] to 12 hits, all in the same place, to inflict major damages on many of the imposing Japanese fortifications (VAC ArtpRpt 29).
>
> Like air and naval gunfire, artillery was severely handicapped by limited observation. Forward observers could seldom see more than 200 yards to the front, and in the more rugged areas, particularly in the 4th Division zone of action, this visibility was often reduced to 100 yards. As a result, although forward observers controlled the majority of the missions, a greater dependence had to be placed on airborne spotters, especially for counter-battery and destruction missions.[4]
>
> — *Iwo Jima: Amphibious Epic*
> Historical Branch, HQMC

The enemy, on the other hand, had been operating from a position of strength. Not only could he still see most targets from his remaining hilltops, but his guns were of a heavy enough caliber to damage U.S. armor.

> For the first week of the drive north, the Japanese on Iwo actually had the attacking Marines outgunned. Japanese 150mm howitzers and 120mm mortars were superior to most of the weapons of the landing force. The Marines found the enemy direct-fire weapons to be equally deadly, especially the dual-purpose antiaircraft guns and the 47mm tank guns, buried and camouflaged up to their turrets.[5]
>
> — *Closing In: Marines in the Seizure of Iwo Jima*
> History and Museums Division, HQMC

In effect, many of the Japanese gunners had been using their big cannons like sniper rifles. But, being outgunned—once acknowledged by those in charge—would help to save Operation Detachment from disaster. Division leaders soon realized that safely taking an attack objective had more to do with the maneuver of tiny infantry elements than with overwhelming firepower.

> Many cleverly camouflaged positions escaped discovery until advancing troops were so near that heavy supporting weapons could not be used against them. Others, although observed and taken under fire, withstood even the largest shells available on Iwo, and could be silenced only by combined tank-engineer-infantry action.[6]
>
> — *Iwo Jima: Amphibious Epic*
> Historical Branch, HQMC

What Had Seemed Relatively New to the Marines on Iwo

The Marine infantrymen who had hit Iwo were theoretically prepared to penetrate a hardened defense line.

> Every intricate detail of wiping out pillboxes, concrete emplacements and heavily fortified positions had been studied and rehearsed.[7]
>
> —*Third Marine Division,* by Aurthur and Colhmia

But, the particulars of that training have since been lost to history. Later generations suspect there had been more emphasis on how to silence a bunker than sneak up on one. Most important to its neutralization was getting an explosive through the aperture. For that, the Marines would sometimes require a pole charge. Unfortunately, most blockhouses had been so well constructed that even after their frontal chambers had been engulfed by flame or concussion, their guns could be quickly remanned.

> The positions were masterpieces of concealment and construction: walls of many were more than three feet of steel-reinforced concrete and impossible to spot from the natural terrain upheavals that camouflaged them. Some measured ten by twenty feet, had five-foot ceilings, and were made up of three separate rooms with twelve-inch concrete walls, the areas connected by narrow crawl-ways.
>
> . . . Even when Marine demolitions men could fling a satchel charge into a gun vent, the blast would have little or no effect on the other two rooms.[8]
>
> — *Iwo Jima,* by Ross

And, of course, all Japanese blockhouses and gun caves were surrounded by up to 10 machinegun nests. Every Japanese squad of that era had its own Nambu.[9]

Some Parts of the Puzzle Remained Fuzzy

The amount of enemy weaponry to survive the heavy U.S. bombardment came as a shock to the island's new occupants. Iwo may have been the first of the Marines' Pacific battles where their maneuver forces and resupply points/convoys would remain under constant surveillance by enemy supporting-arms observers. Suribachi was 554 feet in elevation. After that dormant volcano had been taken, all Marine drives onto the central plateau could be seen from six hills over 330 feet high, and three more over 200 feet. (Map 1.1 only showed the main artillery-observation points.)

After all the U.S. barrages, Iwo had become largely devoid of foliage. Every square yard to the front of any defense line could now be covered by enemy fire. And Marine attackers found their own "fire superiority" almost impossible to establish during any push.

Initial Supply Problems

Nowhere in the Pacific had the Marines faced fire as intense as in Iwo's initial ambush. On the way to Airfield No. 1, Marine units had nearly run out of machinegun ammunition. Then, a number of men went running back to the beach for more. By some accounts, that's how Gy.Sgt. John Basilone was killed. He had, after all, been a machinegun section leader. His Navy Cross citation lists the location and cause of death as Airfield No. 1 from enemy mortar fire.[10] Tony Stein would make eight trips back to the beach for more ammunition for his "Stinger." PFC Ralph Edwards was a replacement who assisted a gun team from I/3/24 by repeatedly returning to the beach for bullets.[11]

The Marine assault units had likely planned on vehicles to bring up their ammunition, food, and water, before evacuating the wounded. 380 LVTs (Landing Vehicle Tracked) had hit the beach with troops on board.[12] After most of the wheeled vehicles got stuck, beachmasters were forced to use LVTS and DUKWs to keep the cargo flowing inland. The LVTs could more easily navigate the soft beach and steep terrace. After some were lost to mines or artillery fire,[13] "weasels [tiny tracked vehicles] handled much of the load until around D+4." That's when the roads became durable enough for wheeled traffic.[14] But, up near the Japanese defense lines, long-range enemy shelling remained so intense that "after-dark working-party patrols" were the only way to transport the much-needed cargo.

> [Near the Amphitheater,] infantry carrying parties worked well into the night to bring replenishment supplies forward across fire-swept areas.[15]
>
> — *Iwo Jima: Amphibious Epic*
> Historical Branch, HQMC

Evacuation of the Wounded

Nor was timely medical attention an automatic for injured Marines. Tony Stein had taken a "walking wounded" every time he went back to the beach for ammo.[16] But, the only thing that had saved him and his new friend from being shelled was more lucrative targets.

Sadly, not all of those wounded Marines would make it to a hospital ship. Inexperienced Landing Ship Tank (LST) crews would sometimes fail to lower their bow ramps enough to accommodate DUKS or LVTs approaching them after dark. Many of those amphibious vehicles—though loaded with casualties—became disoriented, ran out of gasoline, and sank. (See Figure 8.1.) In all, 148 LVTs would be lost at Iwo, with 34 falling victim to the surf, and another 88 sinking in deep water (mostly at night).[17] Nor was every casualty lucky enough to make it onto a LST-bound vehicle.

Figure 8.1: LVTs in Heavy Seas Around an LST
(Source: "Military Transport Boats at Iwo Jima" by Harry Reeks, ©, Anne S.K. Brown Collection, Brown Univ, Library.)

Continuing Dangers at the Beach

Iwo's southern beaches would have to endure "incoming" throughout much of the 36-day ordeal. This created considerable danger for the shore party (and its temporarily included replacement draft [18]). Many of the D-Day casualties would also be injured a second time as they lay helpless on the beach. Over the course of the invasion, 1350 Marines and Navy corpsmen would somehow die of their wounds.[19]

> [26 February] . . . .The beach was under mortar fire and every now and then when rounds started falling close, the staff would leave to get in foxholes or trenches. The patients just had to stay in place and hope they didn't get hit. . . . [P]ast midnight, I [Rowland Lewis] was . . . put on an LCM [Landing Craft Mechanized] . . . . We went out . . . and tied up alongside an LST. . . . We finally arrived at the President Adams, a former passenger ship which . . . was now being loaded with casualties. . . . [A] couple of sailors from the working party carried me down to a makeshift hospital ward. It was dawn and it had been about 18 to 20 hours ago I had been hit [in the arm and groin].[20]
>
> — C/1/23 public domain website

Attrition Warfare Has Some Inherent Risks

To take advantage of all available firepower, frontline U.S. units tried to stay roughly on line and tied in with each other. When one got a little ahead of another, the forward-most unit often pulled back a little. If heavily engaged at the time, it could not always account for all members. That could create quite a problem when more artillery fire became necessary.

Upon crossing an Airfield No. 2 runway at the MLR's center, C/1/23 left a whole platoon behind while preparing for additional preparatory bombardment. Whether that friendly shelling ever took place is not known, only that some of the missing Marines would later be found wounded in the likely impact zone.

> We pulled back [across the Airfield No. 2 runway] to about the same position we had occupied . . . a couple or three

hours earlier. . . . Some time later, Capt. McDaniel told me that the third platoon had not come out and I should take a couple of men and go bring them back. . . . I took Herb Johnson and one other, . . . and we went back across the runway. . . . We angled to the right . . . thinking we would be just behind the [missing] third platoon. . . . Of course, that was [also] where there were a lot of Japanese soldiers, but no one from the third platoon. We came under fire immediately. I was hit in the left forearm and . . . upper abdomen. . . . I next saw Herb on a shipload of wounded bound from Saipan to . . . Hawaii. Herb had been hit in the head . . . and didn't know . . . what had happened. . . . I don't know what happened to [the third person].

Figure 8.2: Memories of Guadalcanal
(Source: "Bloody Ridge," by Howard Brodie, Army newspaper artist, in "The Legend of Suicide Charley," by Cozzens, "Marine Corps History," Summer 2017, p. 30.)

> I would judge it was about noon when I got hit. So, I found myself lying along and in shock at the bottom of a shell hole. . . . About mid-afternoon a Navy corpsman . . . put a couple of bandages on me and left. . . . At dusk just as I came out of one of my dreamy periods, . . . I looked up . . . and it was Ed Rajkowski with several others carrying someone in a poncho. Ed was in the first platoon, but when the gap opened he ended up with the third platoon. They were carrying the third platoon leader who had a broken leg.[21]
>
> — C/1/23 public domain website

Of course, the low-ranking Marine volunteers who went out ahead of parent-units looking for enemy machinegun nests were also at extreme risk from friendly fire. More than one may have been assumed dead to allow for a pre-assault bombardment.

The Dehumanizing Effect of 2GW

John Basilone had been forced to gun down hundreds of enemy soldiers on Guadalcanal.[22] (See Figure 8.2.) One photo of him receiving the MoH in Australia reveals a wild sideways glance.[23] "Manila John" had been the regional boxing champ while serving with the U.S. Army in the Philippines from 1934 to 1937,[24] so he was a pretty tough guy. But, he was also a religious and big-hearted guy. He may not have liked what 2GW had been doing to his humanity. Before "staying with his boys" *en route* to Iwo Jima, he had told his family he wasn't coming back.[25]

If, Basilone's initial bunker take-down on Iwo had involved personally shooting 8-10 escaping occupants,[26] that action may have proven a little disconcerting to a straight-up fighter. Then, as his unit moved inland, he returned to the beach for ammunition. All this extra running around would have put him at great risk from Kuribayashi's barrage. Then, shrapnel from a "big-mortar round" finally tore into his abdomen near Airfield No. 1. The same explosion killed four of his men. Big mortar rounds can be heard coming in. Someone with Basilone's level of experience should have been able to dodge all but a direct hit. But, after demonstrating how important frontline initiative was to speeding up 2GW, the bigger-than-life SNCO may have wanted to stay with his dying boys. There exists an old naval tradition of the captain always going down with his ship. And a truly brave man would have had no trouble doing so. The attending corpsman would never reveal Gunny John's final sentiments.[27]

Part Four

Tactical Details of What Happened

"The cost to us in [a] quality,
one who did not fight side by side
with those who fell can never understand."
— Maj.Gen. Graves B. Erskine USMC
(3rd Marine Division Commander on Iwo Jima)

(Source: Maj.Gen. Erskine, as quoted in "The Third Marine Division,' by Aurthur and Cohlmia; ed. by Vance, Washington, DC: :Infantry Journal Press, 1948.)

9 Up to the Flag-Raising on Suribachi

- Did any new U.S. tactics occur almost at once on Iwo?
- Which enemy initiatives had prompted them?

Separate riflemen had no trouble getting involved.

(Source: "Us Marines," painting by Charles Waterhouse, "Leatherneck Magazine," August 1985, cover art, ©.)

The Initial Beachhead

Gen. Kuribayashi had wanted to make the landing as costly for the Americans as possible. To fill the beach with more targets, he had delayed most defensive fires for a full hour. Unbeknownst to U.S. war planners, that beach itself had been a natural trap—so soft and granular that neither man nor tracked vehicle could easily leave it. Nor was it suitable for scraping out a protective hole. That viscous morass of volcanic debris may have absorbed some of the concussion from incoming shells, but it did very little to stop their shrapnel. (See Map 9.1.)

Yet, Iwo's shore had not been totally free of small-arms fire. Some of the previously damaged pillboxes became suddenly full of machinegunners from below. That's when John Basilone showed everyone how most safely to counter such a threat—with one or two men instead of a whole unit on line.[1] A PFC on the next wave into an adjacent beach may have witnessed the famous Gunny's example and then tried it himself.

> 19 . . . February 1945. Advancing on a heavily fortified Japanese blockhouse after landing on the fire-swept beachhead on D-Day, Private First Class Bell [of I/3/23] gained the entrance of the emplacement against a shattering barrage of machine-gun fire and, launching a daring lone attack with grenades, succeeded in destroying a machine gun and several of the enemy therein. Unable to reduce the intricate construction alone, he again exposed himself to devastating fire to direct a flame thrower tank in completing its destruction. When his platoon leader became a casualty during a subsequent bitter action, he promptly assumed command and, after reorganizing his men, advanced with a companion to destroy a machine-gun nest which had been inflicting heavy casualties and retarding the advance of his platoon. Continuing his daring tactics when machine-gun fire emanating from a cave to the front of the line halted the advance, he again volunteered to rush the position and, hurling his grenades with deadly accuracy, assisted in killing three of the Japanese and in destroying a machine gun.[2]
>
> — Navy Cross citation for PFC Joshua S. Bell

There had been no man-made obstacles on the beach, just the soft black sand and difficult-to-climb lava particle terrace.

> Very little barbed wire was encountered; not because barbed wire was omitted as part of the planned defenses but, according to a POW, Major Kara, because of its scarcity due to ship sinkings.[3]
>
> — *Intelligence Report,* G-2, TF 56, April 1945

Just beyond the terrace lay other problems—most notably,

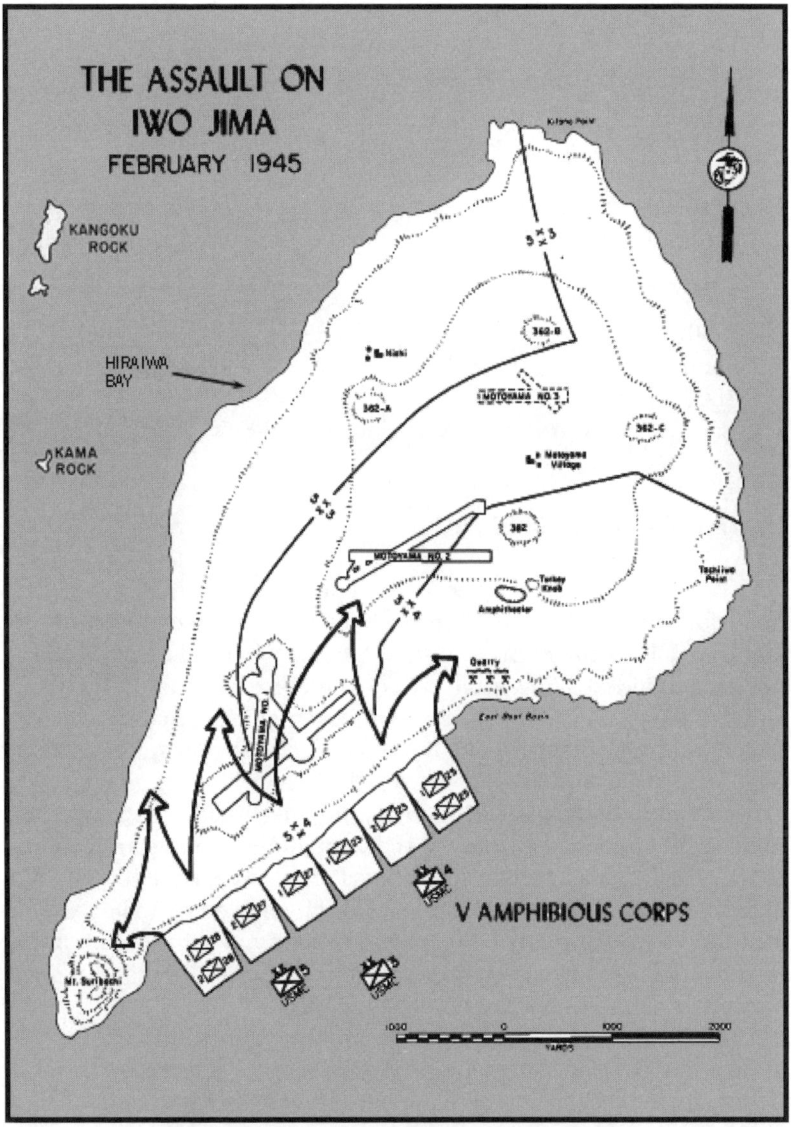

Map 9.1: Landing Plan for Iwo Jima
(Source: Alexander, "Closing In," Marines in WWII Commemorative Series, Washington, D.C.: History and Museums Division, HQMC, 1994, map, page 13.)

antitank trenches, minefields and grazing machinegun fire. Soon, all Marines trying to negotiate them would come under the same barrage of large rockets and shells.

> As of 10 Feb 45, . . . there were:
> Personnel trenches - 42,100 yards (approx.)
> Anti-tank trenches - 6,700 yard's (approx.)
> The bulk of these trenches are found in the southern half of the island, along and behind the beach areas.
> Wire. Approximately 1,600 yards of [presumably barbed] wire were observed on photographs to 10 Feb 45, chiefly in the northeastern part of the island [near *Ginmeisui]*.[4]
> — *Intelligence Report,* G-2, TF 56, April 1945

> [320mm] spigot mortars [could shoot] 1440 yds.; rocket bombs [containing] 310 lbs of explosive [reached as far as] 7500 yds.[5]
> — *Intelligence Report,* G-2, TF 56, April 1945

On the Way to Airfield No. 1

Gen. Kuribayashi had correctly guessed the Yanks would first head for Airfield No. 1. That's where he would spring the final, small-arms portion of his ambush. The nearly level airport offered no place to hide. Lying squarely between Suribachi and Motoyama defense zones, and with rear-side emplacements to oppose a western beach landing, its entire expanse could be easily swept by grazing fire.

To complete the massacre, many Japanese soldiers the Marines had fought or dodged *en route* to the airstrip then reappeared behind them. The isthmus was the lowest part of the island and its soil often of higher-temperature. Yet, occupants of its machinegun nests could still hide beneath them.[6] So, on the first day ashore, the invading force would face automatic-weapons fire from all four sides at once, as well as the indirect-fire bombardment.

But, Kuribayashi had underestimated the frontline Marines' ability to think for themselves. Before long, a disturbing trend in U.S. tactical method began to appear. It was the same type of thing Gy.Sgt. Basilone had demonstrated—for all to see—at the beach. Others had simply followed his example.

> 19 February 1945. Assailed by a tremendous volume of small-arms, mortar and artillery fire as he advanced with one squad of his section in the initial assault wave, Sergeant Cole [of B/1/23] boldly led his men up the sloping beach toward Airfield Number One. . . . Sergeant Cole, armed solely with a pistol and one grenade, coolly advanced alone to the hostile pillboxes. Hurling his one grenade at the enemy in sudden, swift attack, he quickly withdrew, returned to his own lines for additional grenades and again advanced, attacked, and withdrew.[7]
>
> — MoH citation for Sgt. Darrell Samuel Cole

Before dark, the Marines finally managed to cross the isthmus and secure part of Airfield No. 1, but not without cost. Including those lost at the shoreline, the U.S. casualty total for the first day would be over 2400. While the defenders of Suribachi had been cut off from all support, the Marines would need four more days to get close enough to climb it.

Events at the Right End of the Landing Beach

Some of the bitterest fighting of the initial period ashore would occur on the high ground above the Quarry. 3/25 at been the unit to debark closest to those gun-studded cliffs. It and 1/25 would then embark on an epic struggle to knock out the foe's heavy guns.

> There was no cover from enemy fire. Japs dug in[to] reinforced concrete pillboxes laid down interlocking bands of fire. . . . Camouflage hid all enemy positions. The high ground on either side was honeycombed with layer after layer of Jap emplacements. . . . Their observation was perfect; whenever a Marine made a move, the Japs would smother the area in a murderous blanket of fire.[8]
>
> – *Closing In: Marines in the Seizure of Iwo Jima*
> History and Museums Division, HQMC

As had Basilone and the NCOs of other units, it was now up to the enlisted leaders of 4th Division to set the example.

> [O]n 19 February . . . . When his platoon received many

> casualties and the advance was held up by destructive fire from an anti-boat gun emplaced in a cave, Sergeant Doiron [Plt. Guide of K/3/25] immediately secured a bazooka and, moving to an exposed position directly in front of the cave, destroyed the gun and annihilated its crew.[9]
>
> — Navy Cross citation for Sgt. Camille J. Doiron

Because the severity of the fighting on this part of the battlefield, an "after-hours" foray seemed the most logical thing for another NCO to do. (Only much later would a whole U.S. unit risk exploiting the shadows before dawn.)

> 19 February 1945. . . . Corporal Sivicky [of I/3/25] voluntarily entered enemy territory under cover of darkness to place demolition charges in . . . the [enemy] emplacement.[10]
>
> — Silver Star citation for Cpl. Edward J. Sivicky

It wasn't long before a nonrate from a right-side battalion would take matters into his own hands. To stop the foe's direct fire from reaching the waterline, he conducted a one-man envelopment.

> 19 February 1945. . . Private First Class Vincent [of L/3/25] led his squad in a determined assault against enemy coastal emplacements along the fiercely-defended shores of Iwo Jima. . . . Private First Class Vincent, skillfully maneuvering to a position behind a Japanese pillbox while under the protective fire of his team, attacked and killed the hostile gun crew before he himself was mortally wounded.[11]
>
> — Navy Cross citation for PFC James K. Vincent

There was a reason the fighting had been particularly intense above the right end of the landing zone. The enemy's first cross-island defense line had been anchored there.[12] That means the likely reinforcement or resupply by tunnel of any Japanese emplacement under attack.

1/28 Had Come Across the Beach Closest to the Volcano

Plunging fire from Suribachi had been equally devastating to the units at the left end of the landing zone. Every move of 1/28 could be contested that way. Additional fire was coming from closer, ground level emplacements. But, once again, the NCOs wasted no time attacking its source.

> 19 February 1945. Landing on the heavily fortified beachhead during fierce hostilities on D-Day, Corporal Shaker [of 1/28] fearlessly risked his life in a valiant attempt to wipe out a 20mm. anti-aircraft gun battery situated on commanding ground and firing relentlessly on units of his battalion. Courageously advancing alone under the enemy's fierce, concentrated shell fire, he boldly charged the emplacement with grenades and fixed bayonet and, fighting furiously against tremendous odds, succeeded in killing the 12 Japanese troops manning the weapon before he himself was mortally wounded.[13]
>
> — Navy Cross citation for Cpl. Conrad F. Shaker

But Corporal Shaker was not the only Marine from 1/28 with initiative. That battalion also had a former armorer who had figured out a way to more easily approach a bunker.

> 19 February 1945. . . . The first man of his unit [A/1/28] to be on station after hitting the beach in the initial assault, Cpl. Stein, armed with a personally improvised aircraft-type weapon, provided rapid covering fire as the remainder of his platoon attempted to move into position. When his comrades were stalled by a concentrated machinegun and mortar barrage, he gallantly stood upright and exposed himself to the enemy's view, thereby drawing the hostile fire to his own person and enabling him to observe the location of the furiously blazing hostile guns. Determined to neutralize the strategically placed weapons, he boldly charged the enemy pillboxes 1 by 1 and succeeded in killing 20 of the enemy during the furious single-handed assault. Cool and courageous under the merciless hail of exploding shells and bullets which fell on all sides, he continued to deliver the fire of his skillfully improvised weapon at a tremendous rate of speed which rapidly exhausted his ammunition.[14]
>
> — MoH citation for Cpl. Tony Stein

> Stein spotted a pillbox holding up the advance and went for it. His weapon was a one-of-a-kind machinegun fired from the hip....
>
> The husky corporal called it his "stinger," and it was [quite] unique. . . . Spewing bullets in rapid bursts, the "stinger" and its gung-ho triggerman [Stein] pinned the Japanese inside the pillbox while [Sgt. Merritt M.] Savage finished them off with a demolitions charge.[15]
>
> — *Iwo Jima,* by Ross

When 3/28 was summoned to help with the taking of Mount Suribachi, some of its nonrates decided to get involved. They soon discovered that a single attacker could more easily get things accomplished.

> [O]n 19 and 20 February 1945. Separated from his squad upon landing on D-Day and working entirely upon his own initiative, Private First Class Goff [G/3/28 Demolitions Man] destroyed two hostile pillboxes with the limited explosives he carried. Although painfully wounded while destroying the second pillbox, he . . . continued with the squad in the assault. When two more enemy pillboxes halted his platoon's advance, he crawled toward them amidst heavy hostile small-arms fire and, after throwing several hand grenades through the fire ports of the first one, placed a demolition charge which completely destroyed the emplacement. Again wounded while repeating the performance on the second pillbox, he remained long enough to place a charge and ignite it, thereby destroying the pillbox.[16]
>
> — Navy Cross citation for PFC William A. Goff

How the Base of Surbachi Was Eventually Taken

Suribachi had been covered with hundreds of spider holes and gun caves. Yet, it was never stormed *en mass* from below as was the custom in those days. Instead a platoon-sized patrol would fight its way up the tiny volcano—through a maze of enemy positions—and plant a U.S. flag at its top. Consolidation of the mountain would then ensue—from top down. But, none of this would be possible until the base of the mountain was reached.

> Some of Suribachi's stoutest defenses existed down low, around the rubble-strewn base. Here nearly 70 camouflaged concrete blockhouses protected the approaches to the mountain; another 50 bulged from the slopes within the first hundred feet of elevation. Then came the caves, the first of hundreds the Marines would face on Iwo Jima.[17]
>
> – *Closing In: Marines in the Seizure of Iwo Jima*
> History and Museums Division, HQMC

Again the single-handed way of doing things took center stage, this time from a Corporal.

> 21 February 1945. . . . Assigned with his squad to blast a series of mutually supporting caves retarding the advance of Landing Team 1/28 at the base of Mount Suribachi on D-Day plus 2, Corporal Bahring voluntarily undertook the perilous mission himself and, boldly advancing along the broken, jagged terrain, conducted an accurate reconnaissance to locate the emplacements. Constantly the target of hostile grenade and small-arms fire, he courageously edged his way up the incline and, single-handedly placed his charges, closed each cave in turn and enabled the infantry to renew its advance up the mountain.[18]
>
> —Navy Cross citation for Cpl.Raymond W.Bahring

After finally realizing what it took to make any forward progress, a few young officers then ventured forth to better direct the supporting arms.

> [O]n 20 and 21 February 1945. . . . Defying uninterrupted blasts of Japanese artillery, mortar, rifle and machine-gun fire, Captain Dunlap [of C/1/26] led his troops in a determined advance from low ground uphill toward the steep cliffs from which the enemy poured a devastating rain of shrapnel and bullets, steadily inching forward until the tremendous volume of enemy fire from the caves located high to his front temporarily halted his progress. Determined not to yield, he crawled alone approximately 200 yards forward of his front lines, took observation at the base of the cliff 50 yards from Japanese lines, located the enemy gun position and returned to his own lines where he relayed the vital

> information to supporting artillery and naval gunfire units. Persistently disregarding his own personal safety, he placed himself in an exposed vantage point to direct more accurately the supporting fire and, working without respite for two days and two nights under constant enemy fire, skillfully directed a smashing bombardment against the almost impregnable Japanese positions despite numerous obstacles and heavy Marine casualties.[19]
>
> — MoH citation for Capt. Robert Hugo Dunlap

Soon, low-ranking engineers noticed that no amount of U.S. shelling was able to silence the guns in those caves. One must have been a rock-climbing enthusiast in civilian life.

> 22 February 1945. When heavy fire from an enemy cave at the base of Mount Suribachi disrupted the advance of his company and harassed our men in the rear areas, Private First Class Babich [engineer with an infantry company] volunteered to go ahead of the Infantry lines and try to silence the emplacement. Undaunted by the fact that the cave was in the walls of a cliff forty feet high and that he was exposed to heavy hostile fire from all directions, he approached the base of the cliff and cautiously climbed to the top. Despite the extreme danger, he carefully placed a heavy demolition charge over the cave and returned to his original position. Discovering that the charge had a defective fuse, he repeated his first daring mission, setting off a second charge which sealed the mouth of the cave and trapped the Japanese inside.[20]
>
> — Navy Cross citation for PFC John B. Babich

Up Suribachi

At D+4, the executive officer of the same company in which 1stLt. Wells had commanded 3rd Platoon, finally led that platoon up Suribachi.[21] His name was Schrier. In single file, the 50 or so men of that reinforced platoon followed an old Japanese trail up the north side of the partially collapsed volcano. They climbed part of the way under the protective fires of their battalion.

> On the morning of 23 February when his combat team had advanced to the base of Mount Suribachi after four days of severe fighting, First Lieutenant Schrier [of E/2/28] volunteered to lead a forty-man patrol up the steep slopes of the mountain. Quickly organizing his patrol and placing himself at its head, he began the torturous climb up the side of the volcano, followed by his patrol in single file. Employing the only known approach, an old Japanese trail, he swiftly pushed on until, covered by all the supporting weapons of his battalion, he gained the top of the mountain despite hostile small-arms and artillery fire. Forced to engaged the remaining enemy in a sharp fire fight, he overcame them without loss in his patrol and occupied the rim of the volcano.[22]
>
> —Navy Cross citation for 1st Lt. Harold G .Schrier

When the Stars and Stripes finally flew, its importance to America's greatest generation became painfully clear. After being unmercifully pummeled from above for four days, "strong men wept unashamedly."[23] Never again would they witness a U.S. flag flying in a stiff breeze without strong feelings of emotion. An injured Wells would return to his platoon after that flag was raised, but then be ordered to relinquish command to Sgt. Ernest "Boots" Thomas. Killed in action several days later,[24] Sgt. Thomas may have been the inspiration for Sgt. Stryker in "Sands of Iwo Jima."[25] Or, it may have been Sgt. Strank—a former Raider who helped bring the second flag up the hill and was killed on 1 March.[26] Either way, another legend would be born.

10 Through the Second Defense Belt

- Which individual ploys paved the way for unit movement?
- What weapons were employed?

Individual actions were the key to MLR penetration.

(Source: Charles Waterhouse painting, part of "Closing In," by Col. Alexander, Marines in WWII Commemorative Series, HQMC, 1994 pamphlet.)

Where Was the Second Defense Line?

The Second Defense Line would turn out to be the Japanese MLR. But, its exact trace had not been evident from aerial photographs before the invasion. Its supposed location on Map 1.1 would turn out to be only partially correct. Instead of a linear string of strongpoints, it would turn out to be a wide belt of strongpoint matrices encompassing all nearby high ground, to include Hills 199-Oboe and Peter. Much of this belt would be composed of tiny ridgeline fortresses that were indistinguishable from the surrounding terrain. But trying to breach it at any point drew a veritable blanket of mutually supporting enemy fires. This chapter will hopefully ascertain the belt's precise path, as well as how U.S. forces finally managed to attack along it.

Some Ammunition Had Become Hard to Come By

With the Japanese still able to target any place on the island with indirect fire, the American divisions had headed northeastward. Many of the vehicles on which they had counted for resupply were already disabled. Others would be almost certainly hit trying to reach the frontline units. But, the U.S. style of warfare had counted on fire superiority over the immediate opposition to move forward. So, at the front, some U.S. ordnance became quite scarce. This shortage may have helped Marine commanders to finally realize that firepower and surprise are interchangeable in combat,[1] and that the only sure way to silence a dug-in enemy machinegun is to sneak up on it. Meanwhile, their enemy counterparts—who were already practicing this advanced style of warfare (3GW)—kept doing all they could to limit the U.S. flow of ammunition. They were mostly relying on indirect fire and infiltrators.

> On one occasion, well-aimed Japanese fire detonated the entire Fifth Marine Division ammo dump. In another tense moment, the ammunition ship Columbia Victory came under direct Japanese fire as she approached the western beaches to commence unloading. . . . The ship escaped, but the potential still existed for a disaster of catastrophic proportions.[2]
>
> – *Closing In: Marines in the Seizure of Iwo Jima*
> History and Museums Division, HQMC

Saving the Colombia Victory had all the earmarks of Providential intercession. But, the Americans' eventual success on that hellish little island was still far from certain. The enemy's new 3GW-oriented defense had been evading most of the frontal U.S. fire. His pillboxes were protected by careful camouflage, "safe rooms," and backside exits. His ridgeline machinegunners could shift via nearby tunnel to reverse-slope positions. And his spoiling attacks could be run through passageways beneath American lines.

Foe's First Defense Line Lay Just Beyond Airfield No. 1

Map 1.1 has depicted the assumed location of that initial strongpoint string at the time of the invasion. Subsequent events were to prove that estimate to be fairly accurate. Airfield No. 1 sat atop a small rise studded with tiny machinegun bunkers. And, just beyond the airfield, the Marines would run into heavier resistance at the very edge of the plateau. Many of those plateau guns had shot at them from a flank as they crossed the isthmus.

> [T]he [first] cross-island defensive system . . . was located on the high ground behind Airfield No. 1. These organized positions extended from the north end of the southeast beaches, across the island to the general vicinity of Nishi [Village and Hiraiwa Bay].[3]
>
> – *Intelligence Report,* G-2, TF 56, April 1945

First Defense Line Ran Past Airfield No. 1 to the Boat Basin

While the principal inland strongpoints had been carefully disguised to evade pre-assault bombardment, landing-force planners had obtained an enemy intercept containing the location of the First Defense Line.

> [A] captured [Japanese] document, dated 10 July, 1944, stated: "A part of the [2nd Mixed] Brigade will defend the east and west beaches, but the main strength will construct positions extending from *Nishi* Village through the area north of the old naval guard barracks to the *Minami* boat landing [possibly #33 on Japanese Map 10.1].[4]
>
> – *Intelligence Report,* G-2, TF 56, April 1945

Enemy landmines were then to provide another clue as to that defense line's exact trace. Such a minefield existed just above the NW-SE runway of Airfield No. 1, and another between there and the eastern shore. According to an official battle chronicle, these man-made barriers had augmented a few that were natural. "[There existed] an intricate network of mutually supporting pillboxes emplaced on high ground between Airfields 1 and 2. . . . The [same] ground afforded the enemy natural tank barriers."[5]

Where Was the Second Defense Line?

That captured document had also described the Second Defense Line. The TA (Target Area) codes in parentheses refer to places on the G-2's map. All information in brackets comes from a Japanese cartographer's rendering after the war.

> Should the above [first-line details] fail, the Brigade will defend the line from *Osaka Yama* [#12 on Map 10.1 or Hill 362-A] through the naval gun emplacements [#17?] to the *Minami* boat landing [#33?] to prevent the enemy's using the airfield [No. 2].
>
> The Nishi (West) Sector was to occupy positions reaching from 500 meters north of Nishi to the No. 2 Airfield. The Minami (South) Sector was to 'occupy' positions extending from southwest of the No 2 Airfield through Minimi Village to the Minami boat landing. The main strength was to occupy positions running from *Osaka Yama* [#12 or Hill 362-A] (. . . TA 216), through *Tamana Yama* [#20 or Navy Headquarters] to *Minami* Village. The main strength was also to construct supplementary positions on the line from the *Osaka Yama* gun emplacement [#12 or Hill 362-A] (. . . TA 216) through the *Motoyama* emplacement [#18 or Hill Peter] to the forward extremity of *Nidan Iwa* [#19 or Hill 382] (TA 201U) and *Tamana Yama* [#20].
>
> Another captured document/undated, of the 2nd Mixed Brigade, stated in part: "Even if finally our main positions should unavoidably fall into the hands of the enemy we will hold the line *Osaka Yama* (TA 216S), *Megane Iwa,* [eastern outcropping of Hill 362-A] (TA 216Y), Navy Gun Position [#17 or Hill 199-Oboe], *Motoyama* Gun Position (TA 200P), *Byobu Iwa* [#21 or part of Charlie-Dog Ridge near airport] (TA 183C), *Tamana Yama* Navy Headquarters (TA 184B).[6]
>
> — Captured document of 2nd Mixed Brigade
> *Intelligence Report,* G-2, TF 56, April 1945

That Post-War Japanese Map

Map 10.1 was probably drawn sometime after the war by a

Japanese researcher. The extent to which it was based on official Japanese documents is unknown. However, its legend still provides credible hints as to what existed at certain locations on Iwo Jima. Comparing the Japanese map to that of the G-2 shows three key spots out of place on the former: (1) Kuribayachi's Cave; (2) *Tamara-yama* Navy Headquarters; and (3) *Higashi-yama* Hill 362-C. The last two turned out to be much closer to the island-long tunnel of the Japanese documentary. In fact, they lay immediately adjacent to it. On the other hand, *Byobu-yama*—which Map 10.1 shows just above that tunnel—may have actually been some 700 meters to the east at Turkey Knob. Whether those inconsistencies were coincidental is anyone's guess. Final events at the tunnel's end may help to solve that mystery.

Role of Iwo's Ridgelines in the Belt

U.S. unit-progress maps fail to agree on where the Second Defense Line was encountered. So, to ascertain its appearance, one must go by the Map 1.1 version and nearby terrain. Progressively taller terraces ringed the northern plateau. Nishi Ridge was part of that shoreline ring. But there was also a series low ridgelines that crisscrossed the island. And, in combination, some must have formed the backbone of the Second Defense Line.

> North of it [Airfield No. 2], lay a series of wild, rocky ridges twisting from shore to shore.[7]
>
> —*Third Marine Division,* by Aurthur and Cohlmia

That precipitous terrain had offered not only a natural barrier to U.S. tanks, but also an elevated view of the battlefield.

> The two runways [of Airfield No. 2] were . . . commanded by high ground several hundred yards to the north. . . . The main hostile battle position . . . apparently extended generally northwest to southeast across the island along the high ground. As long as the Japs held this ridge, they would be able to place observed fire on the Marines' reserve areas and rear installations.[8]
>
> —*Third Marine Division,* by Aurthur and Cohlmia

Just south of Nishi Village lay Hill 362-A. While Map 1.1 fails to say so, this sizeable piece of terrain would prove to be part of the MLR—just as the Mixed Brigade intercept had indicated.

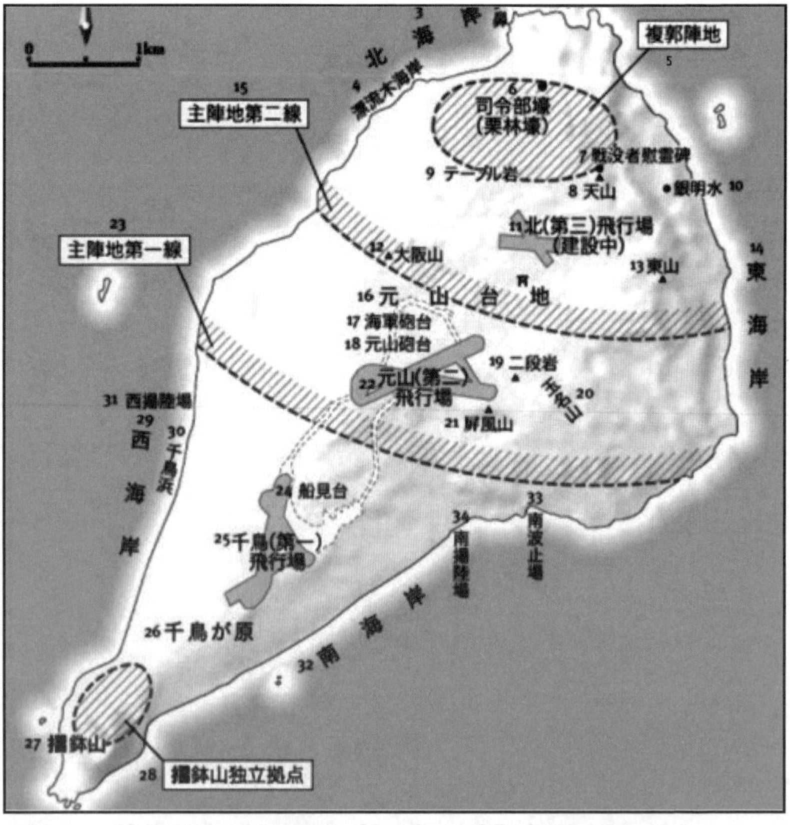

Map 10.1: Japanese Version of Defense Zones

(Source: "So Sad to Fall in Battle," by Kumiko Kakehashi, "©, from url: https://books.google.com/books?id=GTUVD9qN7dQC&printsec=copyright#v=onepage&q&f=false.)

Map Legend

5 *Fukkaku Jinchi* (Honeycomb Defensive Position)
6 Command Bunker (erroneous location for Kuribayashi's Cave)
8 *Tenzan* (Hill 362-B)
9 *Table Iwa* (Hill 291) (109th Infantry Hdqts.)
10 *Ginmeisui* (a fountain from Mt. Fuji in Japan)
12 *Osaka-yama* (Mount Osaka) (Hill 362-A)
13 *Higashi-yama* (Mount Higashi) (erroneous location for Hill 362-C)
14 East shore
15 Second Main Line of Defensive Positions
17 Naval Gun Battery (Hill 199-Oboe)
18 *Motoyama* Gun Battery (Hill Peter)
19 *Nidan Iwa* (double ledge) (Hill 382)
20 *Tamana-yama* (Mount Tamana) (erroneous location of low promontory containing Navy Hdqts.)
21 *Byobu-yama* (Mount Byobu) (vertical ledge) (this low Charlie-Dog Ridge promontory next to the airport is by other accounts Turkey Knob, some 700 yards to the east)
23 First Line of Main Defensive Positions
24 *Funamidai* (part of Airfield No. 1 next to strange dashed-line area leading to Airfield No. 2)
28 *Suribachi* Independent Base
33 Southern Jetty (possibly Minami Boat Landing)
34 Southern Unloading/Disembarkation Point (possibly Minami Boat Landing)

Note: All entries checked against G-2, TF56 Terrain Map

On D plus 9, the 5th Marine Division was squarely up against Hill 362[-A] . . . the backbone of the Japanese defense [line]. . . .

> . . . . 1/27 attacked a jagged, irregular ridge line running more or less from Hill 362[-A] down to the western beaches.[9]
>
> —*WWII History of 5th Marine Division,* by Connor

Farther down that same 362-A ridgeline (on other side of Airfield No. 2) was a section of MLR that was to see enough action to earn itself a nickname—the Meatgrinder.

> Running along the southeast edge of the east-west runway of Airfield Number 2 was a ridge known only as "Charlie-Dog Ridge." In prolongation of this ridge, on a lower level to the southeast, was a semi-circular rise of ground that formed the "Amphitheater." In this area, Japanese engineers had constructed some of the most formidable defenses on the island.[10]
>
> — *Iwo Jima: Amphibious Epic*
> Historical Branch, HQMC

Japanese Strongpoint Mechanics

The average Iwo strongpoint consisted of a pillbox or cave with a semi-circle of machinegun nests to its front. The ridgeline strongpoints often had tunnels to well-camouflaged fallback positions. (Look back at Figure 2.3.) From there, withdrawing occupants could fire at passing Marines from behind, if they chose. Of course, some of those strongpoints had been originally placed on the backsides of ridgelines to protect them from direct U.S. fire. According to the University of Missouri, "The Japanese [had] used reverse-slope defenses with much success in the Philippines, Iwo Jima, and Okinawa campaigns."[11]

> Company K [3/21] had attacked and once more gained a foothold on Hill 199 when suddenly a wave of Jap infantry rose out of a gully on the reverse slope of the hill and counterattacked with bayonets.[12]
>
> —*Third Marine Division,* by Aurthur and Cohlmia

Several fortified ridgelines or flatland bunker strings in succession then constituted an in-depth defensive belt. Because of the greater degree of mutually supporting fires, such belts were extremely hard for the Marines to penetrate. (Look back at Figure 3.1.)

What Other Nipponese Tunnels Made Possible

North of Airfield No. 1 lay the central plateau and several small hills—all comprised of easy-to-excavate volcanic rock. (See Map 10.2.) Because of that rock's softness, all sections of each defense belt had been undoubtedly linked by tunnel. Enemy soldiers could then fight for one strongpoint and shift to another in the same string. Or, they could follow lateral conduits to a perpendicular one that led back to the succeeding belt.

But, those Japanese soldiers did not always have to travel sideways or rearwards. After each belt had been overrun, next-belt occupants could also move forward into previously occupied positions to conduct reconnaissance patrols, scrounging forays, and spoiling attacks.

Initial Action at the First Defense Belt

The most obvious part of the initial belt ran from just north of Airfield No. 1 to the bluffs above the Southern Jetty. As such, its eastern end had been contested within days of the landing.

> 19 February 1945. Landing on D-day with the first wave of infantry, Second Lieutenant Shannon [of A/1/25] courageously led his platoon inland in the face of intense hostile mortar, rocket, artillery and small-arms fire. When his company's right flank was held up by heavy machine-gun fire from an enemy pillbox, he advanced thirty yards ahead of his front line troops with two riflemen covering him and, after failing in an attempt to destroy the Japanese emplacement with hand grenades, courageously led a charge on the two riflemen. Although subjected to devastating hostile machine-gun fire from two mutually supporting enemy positions [indication of a strongpoint matrix] during the charge, Second Lieutenant Shannon continued on inside the position and, when his carbine jammed, clubbed

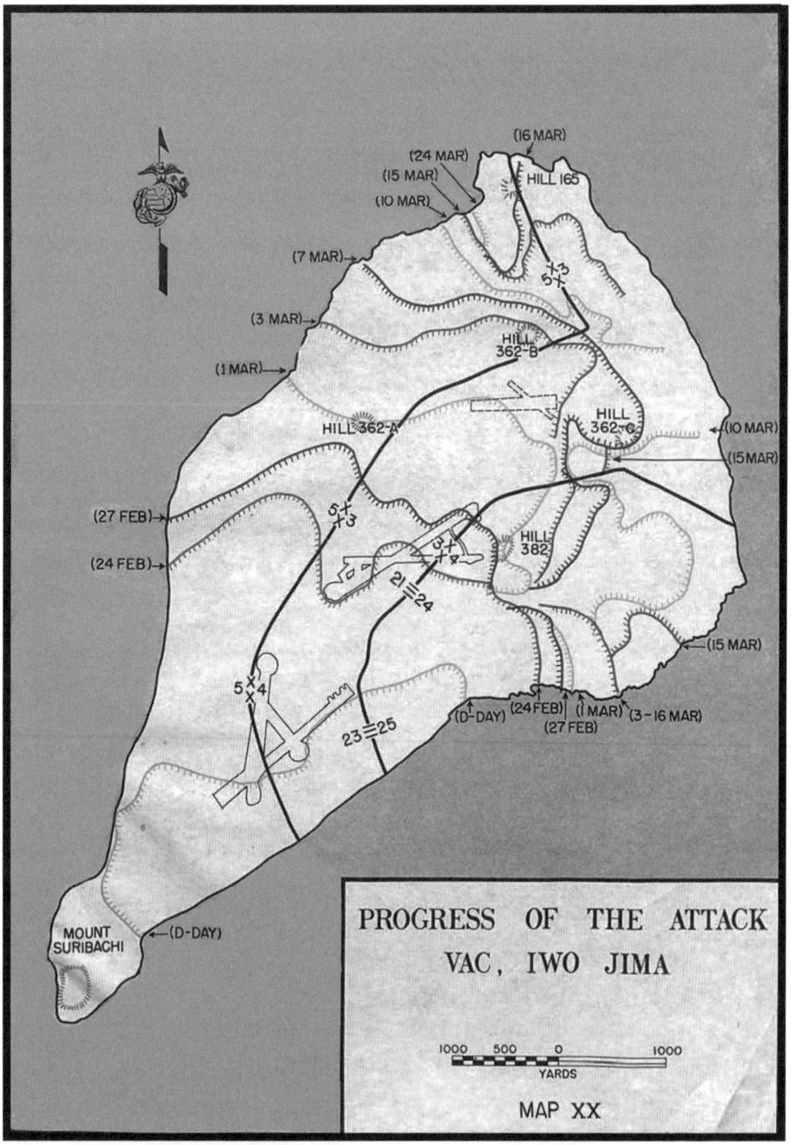

Map 10.2: Divisional Lanes As the Battle Progressed
(Source: Lt.Col. Whitman S. Bartley, "Iwo Jima: Amphibious Epic," Washington, D.C.: History Branch, HQMC, 1954, map XX.)

> the enemy manning the machine gun to death while his two comrades bayoneted the remaining troops.[13]
>
> —Navy Cross citation for 2dLt.William L.Shannon

Though Suribachi had yet to be taken, the first coordinated U.S. offensive to the north began on 21 February with two divisions on line across the entire breadth of the island. The right end of that formation almost immediately encountered the foe's First Defense Belt.

U.S. Units Were Soon Facing Too Much Enemy Fire

At the east end of the First Defense Belt, the terrain had been so rocky and irregular that the Japanese could not block all avenues of approach with minefields. Through such a loophole, U.S. forces could theoretically conduct a tank-supported large-unit maneuver. But, most gaps of this nature had still been preregistered for indirect enemy fire. So, American commanders were forced to do something unusual. That's when incremental small-unit jumps began to replace whole-unit moves—like the Marine Raider fireteams had done at Baikoro.[14] There's no known picture of an online assault by a large U.S. unit on Iwo. While such things may have occurred on the first day while trying to outrun Kuribayashi's firestorm, tiny-element forays soon came to define all big unit advances. They sometimes involved one or two people crawling forward through the rubble to silence an enemy machinegun. But, other times, they involved concurrent small-unit activity in parallel lanes. This would not be the only change to "standard operating procedure" on Iwo. Double envelopments had always run contrary to the Marines' doctrine. Now they were occurring more frequently.

> The 25th Marines moved out again at 0815 on D plus 9 with the same scheme of maneuver as the day before. The 1st Battalion passed through the right flank of RCT 23, with Companies A, C, and I (3/24) in column, to attack Turkey Knob from the north, while Company B stayed on the low ground southwest of the Knob to assault eastward around the base of the Amphitheater to contact other elements of

> the battalion, thus completing a double envelopment.[15]
> — *Iwo Jima: Amphibious Epic*
> Historical Branch, HQMC

Instead of assaulting all in a row, the Marines next started to penetrate enemy defenses in column. That's because columns could more easily change direction and travel a long distance. In the following case, the companies of one battalion attacked in column across the front of another battalion, thereby enjoying a considerable base of fire that required no signals.

> The 21st Marines moved out at 0800 with 1/21 and 3/9 in the assault. The 1st Battalion had moved up from the rear to a position on the right of 2/21, and was to attack in column of companies across that unit's front to clear the rest of Airfield Number 3.[16]
> — *Iwo Jima: Amphibious Epic*
> Historical Branch, HQMC

Tackling the Second Defense Belt

U.S. commanders had yet to confirm that the foe's Second Defense Belt was also his MLR. But G-2 strongly suspected it.

> The main defense line was a belt of mutually supporting positions organized in depth, running generally northwest-southeast across the island. It stretched from the cliffs north of the western beaches south to include Airfield Number 2; then, turning eastward through Minami, terminated at the rugged coast north of the eastern beaches.[17]
> — *Iwo Jima: Amphibious Epic*
> Historical Branch, HQMC

Not until the center of that Second Belt was reached did it become undeniably the MLR. One official battle chronicle describes a "brigade headquarters . . . about 800 yards east of Airfield Number 2 [that] consisted of circular tunnels capable of sheltering 2,000 troops."[18] That massive a complex could only have been along an MLR. It would soon come to be known as the Amphitheater and Turkey Knob.

All three U.S. divisions would be required to take the Second Defense Belt. None breached it and moved on. All made penetrations, and then attacked sideways—into the thickest grouping of strongpoints—to clear their respective sections. To fully investigate the tactics involved, one must first establish that Belt's precise path. That's why the seizure of all prospective bastions will be discussed from north to south, instead of in their chronological order.

Fighting at the Very North End of the Second Belt

An incident in the general vicinity of Nishi Village on 7 March would place where it happened near the top of the Second Defense Belt. 1/26 had set about reducing fortifications holding up its advance the previous day. After little opposition, it reached a 30-foot knoll just north (but some distance to the west) of Nishi Village. As Marine infantrymen surrounded it, enemy fire all but ceased. Demolitions men blasted and closed one cave entrance, and machine-gunners shot several Japanese who came rushing out the back. That's when about 40 jubilant Marines of H/3/26 decided to climb the tiny hill. Then, the unthinkable happened. Its crest blew up with a roar that could be heard for miles. The enemy had just detonated either a CP or ammo dump, thereby inflicting forty-three Marine casualties. All that was left of the tiny rise was a ragged hole in which an apartment building could have fit. Later that day, 3/26 ran into stiffer resistance. So did the 27th Marines on the following day. The only regiment to make any progress was 28th Marines along the coast. There, it encountered very few problems from Hill 215. The left end of the enemy's Second Defense Belt had obviously been crossed. Now, there was only to finish clearing the 5th Division's portion. Hill 362-A to the south (with Japanese code name *Tabaru zaka)* had already been dealt with. Within that formidable bastion were four separate tunnel systems, one of which stretched 1000 yards and had seven entrances.[19] (Look back at Figures 2.6 and 2.7.)

The Seizure of Hill 362-A

If the upper portion of the Second Defense Belt most certainly contained Hill 362-A, just like the 2nd Mixed Brigade document had said. So, that sector ended up looking quite different from its approximation on Map 1.1. 362-A was in the 5th Division's zone, and 3/27 the first battalion to attack it. 3/27 would have difficulty even reaching its base. (See Maps 10.3 and 10.4.)

> 27 February 1945. With casualties increasing steadily under the intensified fire from the enemy's heavy mortar, artillery and infantry weapons on the strongly defended approaches to Hill 362[-A], Gunnery Sergeant Wherry [of G/3/27] fearlessly advanced to the head of the assault units and personally led them against a heavily fortified pillbox which was holding up the right flank of his company and despite the withering barrage succeeding in effecting the destruction of the enemy strong point before he himself lost his life.[20]
>
> — Navy Cross citation for Gy.Sgt. John H. Wherry

> 27 February 1945. With the advance of his company toward Hill 362[-A] disrupted by vicious machine-gun fire from a forward position which guarded the approaches to this key enemy stronghold, [another of G/3/27's SNCOs] Gunnery Sergeant Walsh fearlessly charged at the head of his platoon against the Japanese entrenched on the ridge above him. . . . Thrown back by the enemy's savage resistance, he once again led his men in a seemingly impossible attack up the steep, rocky slope, boldly defiant of the annihilating streams of bullets which saturated the area, and . . . gained the ridge's top only to be subjected to an intense barrage of hand grenades thrown by the remaining Japanese staging a suicidal last stand on the reverse slope.[21]
>
> — MoH citation for GySgt. William G. Walsh

Then, another 3/27 company was met by a Nipponese counterattack.

> 27 February 1945. During an attack on the heavily defended enemy positions guarding the southern approaches to Hill 362[-A], Corporal Towne [of I/3/27] led his men in a successful assault against the commanding ground studded with pillboxes and caves. Severely wounded during a

> furious Japanese counterattack, he continued to direct the fire of his squad by voice and arm signals and, constantly exposing himself to heavy hostile fire, threw hand grenades until he was exhausted and the counterattack repelled.[22]
>
> – Navy Cross citation for Cpl. Harry Towne

The next day, 3/27 was to copy something 2/28 had done on Suribachi.

> The 3d Battalion, 27th Marines . . . by noon [on 28 February] had reached the foot of Hill 362-A. While steady pressure was maintained against this hill fortress, elements of Company H on the right and Company I on the left probed for an opening. Company I pushed a patrol up the southwest slopes to the crest of the hill at 1630 after a day of extremely bitter fighting. But this gain was partially nullified when Company I called in its patrol and dropped back nearly a hundred yards to regain contact with [adjacent] 1/27. . . . Company H then pulled back its right flank about 100 yards to make contact with RCT 21 [of 3rd Division].[23]
>
> – *Iwo Jima: Amphibious Epic*
> Historical Branch, HQMC

After the 362-A crest had been wisely forfeited, 3/27 encountered a really strong counterattack from the direction of Nishi Ridge.

> When his battalion [3/27] was pinned down by intense hostile mortar, machine-gun and grenade fire during a sustained drive to seize heavily fortified Hill 362[-A], on 28 February, Lieutenant Colonel Robertson voluntarily left his position in the forward observation post and, moving along the front line units, inspired his men to heroic effort in resuming the attack until they had advanced up the southern slopes and seized the crest of this vitally strategic hill. With the battalion badly depleted by casualties, only a skeleton platoon left as a reserve, and all regular company commanders and many platoon leaders dead, he rallied his tired, depleted units and directed them to dig in for the night before the enemy, in a desperate attempt to

> split the regimental front, launched a strong counterattack against the left flank of the battalion. Refusing to permit a complete withdrawal, he dashed fearlessly through heavy machine-gun and mortar fire to the imperiled position and ordered an immediate attack which repulsed the Japanese and regained fifty yards of lost terrain.[24]
>
> —Navy Cross citation for Lt.Col.Donn J.Robertson

Of course, the good colonel was not able to accomplish all of this single-handedly. He had the help of his company-grade officers.

> 28 February 1945. With the forward assault elements pinned down by shattering hostile mortar, artillery and small-arms fire during a sustained drive to seize heavily fortified Hill 362[-A], Captain Gray [I/3/27 platoon leader] voluntarily left his position in the observation post and advanced under the withering barrage to the base [forward] platoon [of his company]. After making a quick reconnaissance of the situation, he rallied his weary, depleted units and, despite a serious shrapnel wound received earlier in the action, decided to lead his men in a final, desperate attempt to seize the strong point. With approaches to the objective closely guarded by a fanatic, determined enemy deeply entrenched in rocky ridges and caves, he pushed his relentless advance up the open, gun-studded slopes, stormed the heights of this vitally strategic hill and succeeded in overrunning the position before nightfall.[25]
>
> — Navy Cross citation for Capt. Philip R. Gray

On 1 March, 28th Marine Regiment with 3/27 attached took over responsibility for 362-A from the 27th Marines. By 1030, 1/28 and 2/28 had reached the top of 362-A and the ridgeline running eastward and westward from it. But, they were unable to clear the steep cliff comprising 362-A's northern side. (Look back at Figure 2.6.) The next morning, the two battalions attacked around both ends of the big hill and tried to rejoin. When a company from 2/28 rushed ahead to the next bluff, it suffered heavy casualties. That fire had come from three places: (1) the new bluff; (2) Hill 362-A's north-side cliff; and (3) and a blockhouse to the east. That makes

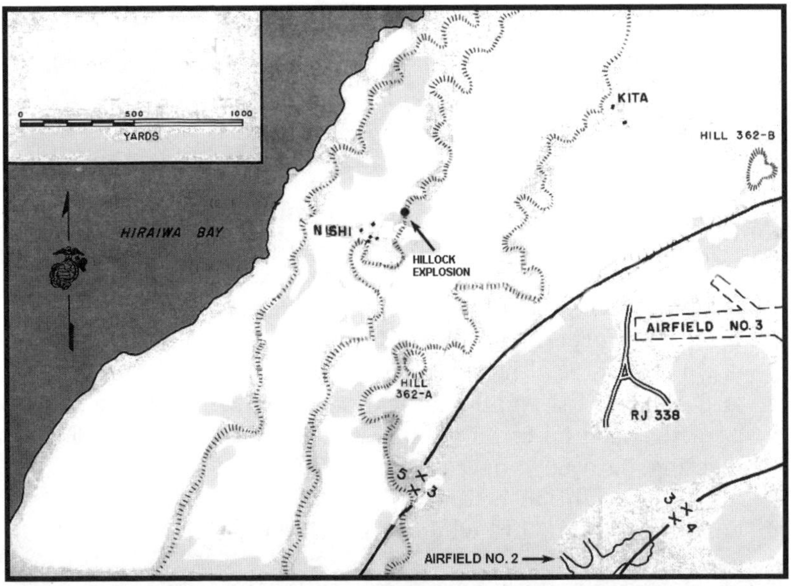

Map 10.3: Terrain and Events at Top of Second Defense Belt
(Source: Lt.Col. Whitman S. Bartley, "Iwo Jima: Amphibious Epic," Washington, D.C.: History Branch, HQMC, 1954, map X.)

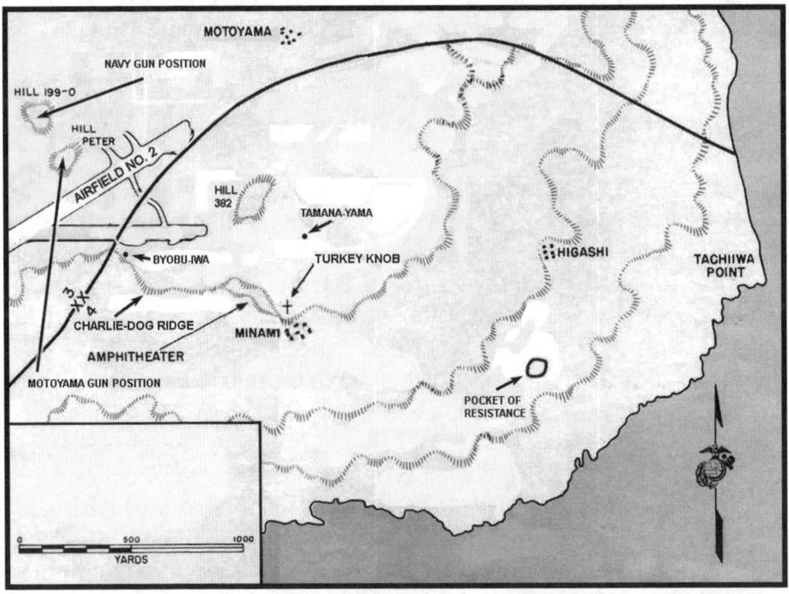

Map 10.4: Terrain and Events at Bottom of Second Defense Belt
(Source: Lt.Col. Whitman S. Bartley, "Iwo Jima: Amphibious Epic," Washington, D.C.: History Branch, HQMC, 1954, maps VIII and XV.)

Hill 362-A no solitary outpost, but a major part of the enemy's MLR. Meanwhile, 1/28 had being having its own troubles on the other side of the hill. The Stinger-toting Cpl. Stein had been just killed while patrolling.[26] Then, another NCO from that same battalion made his ultimate sacrifice.

> With Company B [1/28] pinned down by shattering fire emanating from emplacements in a ravine fifty yards to the front and covering the only line of advance open to the infantry during a drive against [left side of] Hill 362[-A] on 1 March, Corporal Bahring unhesitatingly made his way down the narrow draw through intense machine-gun fire and, hurling his small charges to neutralize the fire temporarily until he could gain an advantageous position for blasting, succeeded in reaching the initial pillbox and placing the demolition before he was fatally struck down by hostile automatic fire while detonating the final charge to destroy the position.[27]
>
> —Navy Cross citation for Cpl.Raymond W.Bahring

Of particular note in this excerpt was the enemy's "draw-following" machinegun fire. U.S. forces almost never defend this way.

Hills 199-Oboe and Peter Just Above Airfield No. 2

Hills 199-Oboe and Peter were vital enough bastions in the defense belt to have Japanese code names. This is again proof of an MLR. The first hill was being called the "Naval Gun Battery," and the second the "*Motoyama* Gun Battery."[28] Anchoring a deep swath of northwesterly running strongpoints, both hills were in the 3rd Marine Division's lane of advance. As with all 2GW battles, U.S. assault troops would soon have more to worry about than enemy bullets.

> [23 or 24 February], K Company [of 3/21] reorganized and drove forward to Hill 199 on the northern edge of the airstrip. . . . Then friendly artillery fire, which had been called down to assist the assault, was resumed through an error and drove K Company back off the hill. Finally the shell-

> ing ceased, and again the attack was launched. Once more the hill was seized, but this time K Company was driven back by the Japs. Both I and K companies were now receiving enfilade fire from the flanks through failure of units on right and left to keep abreast of them. By 1350 Company I had again reached the northern edge of the airfield. Company K had attacked and once more gained a foothold on Hill 199 when suddenly a wave of Jap infantry rose out of a gully on the reverse slope of the hill and counterattacked with bayonets. . . . Later, records showed that 800 pillboxes had defended an area 1,000 yards long by 200 yards deep along that front [that ran from those hills northward].[29]
>
> —*Third Marine Division,* by Aurthur and Cohlmia

The Marines of 3/21 persisted and finally occupied the top of Hill 199-Oboe. But, for how long is not quite clear.

> 23 [or 24] February 1945. Assuming command . . . , First Lieutenant Archambault [of K/3/21] quickly reorganized the company and, personally leading his men in furious hand-to-hand fighting while under hostile frontal, flanking and enfilade fire, succeeded in seizing a strongly fortified Japanese hill position. Counterattacked and driven from the hill three times, he repeatedly led his men against the enemy and finally regained the position and effectively defended it throughout the night.[30]
>
> —Navy Cross citation for 1stLt. R.J. Archambault

Then, on the morning of the 25th, 1/9 and 2/9 took over the fighting for this area. On the left, 2/9 was to head northward from the west end of the airstrip and then pivot eastward against Hill 199-Oboe. On the right, 1/9 was to almost immediately encounter Hill Peter. After a heavy bombardment from naval guns, artillery, and warplanes, their assault got underway at 0930. 1/9 quickly bogged down. Then, nine out of the 26 tanks with 2/9 got hit crossing the end of the runway. But, 2/9 itself managed to move 200 yards ahead of the other battalion. At 1430, 3/9 temporarily filled that gap, but could make no further progress. Still, 9th Marines had seized a foothold on the rising ground north of the airstrip from which to resume the attack. After a heavy U.S. artillery barrage, that push got underway again at 0800 on 27 February. Upon losing another 11 tanks, 2/9 reached the base of Hill 199-Oboe, and 1/9 seized the crest of Hill Peter though partially pinned down by reverse-slope positions. Just as it appeared that no more U.S. gains would be made for the day, the unexpected occurred. Following a short but highly concentrated artillery preparation, 1/9 and 2/9 jumped off again in the late afternoon. This time, the coordination between supporting arms and maneuver unit was perfect. The 2nd Battalion advanced 700 yards. The 1st Battalion overran Hill Peter, continued down its reverse slope, and then drove up the flank of 199-Oboe. Once all fire from these two bastions had been silenced, 2/9 was able to move off to the east quite a way. And, for the first time since the beginning of the offensive, 2/9 and 1/9 were abreast. All of Airfield No. 2 and the commanding terrain just to the north of it were under American control.[31] And the enemy's MLR had been breached at its center.

Mechanics of the Breakthrough

What had just happened? From the severity of the fighting, it's clear the enemy's Main Line of Resistance had been penetrated. A 2GW-oriented attacker would conclude his overwhelming firepower had finally turned the tide. But, why should that last U.S. bombardment have made any more difference than the others? One possible explanation involves the human element. Something may have scared the assault troops more than their battlefield adversary.

> [Around] 27 February 1945. . . . Dauntlessly leading his platoon in repeated attacks against a strong enemy force securely entrenched in caves and pillboxes along a rocky ridge dominating a vital airfield, First Lieutenant Lindblad [of A/1/9] constantly exposed himself to merciless hostile machine-gun, mortar and artillery fire to coordinate the assault and, moving from man to man along the entire platoon front, issued orders and rallied his gallant forces throughout the bitter action. When an intense barrage of Naval gunfire from friendly ships rained down upon his section, disrupting communications, he bravely dashed through the shattering fire for a distance of one hundred

> and fifty yards to report the situation to his Commanding Officer.[32]
>
> — Navy Cross citation for 1stLt. Carl J. Linblad

And then, the real lesson of the action would be what it takes to silence heavily dug-in guns. It first requires infantry penetration of their protective machinegun rings.

> 27 February 1945. During an attack against a strongly fortified hill position when the Platoon Leader and Platoon Sergeant became casualties, Sergeant Villa [of B/1/9] immediately assumed command and, although subjected to intense machine-gun, rifle and grenade fire, led his men in the attack. As the first man to reach the top of the hill, he threw grenades at the enemy fortifications and directed the platoon into commanding positions. When a flame thrower operator was wounded, he reduced two hostile machine gun positions. Then, crawling back to his platoon, he led three demolitionists to newly located Japanese emplacements.[33]
>
> — Navy Cross citation for Sgt. Albert M. Villa

Then, frontline American riflemen had to physically enter the caves housing the big guns.

> 27 February 1945. When his company's advance was held up by strong hostile fortifications, Corporal Vaught [of A/1/9] . . . crawled thirty yards forward in the face of intense enemy machine gun fire and, with his Fire Team covering his advance, neutralized the Japanese emplacement by hurling a hand grenade and annihilating the occupants. Although sustaining a severe chest wound, he aggressively continued forward to a large enemy emplacement housing a coast defense weapon which was delivering fire into our rear areas. He then threw a demolition charge into this position, rendered the weapon inoperable and killed the gun crew.[34]
>
> — Navy Cross citation for Cpl. Orville O. Vaught

Finally, somebody had to completely remove all enemy assets from the offending caves. Such things cannot be accomplished remotely.

> 27 February 1945. When his company was temporarily held up by intense fire from strongly-entrenched enemy machine guns and mortars, Private First Class Fransko [of Hdqts. Co./1/9] dashed across forty yards of fire-swept terrain to gain access to the emplacement. Using his flame thrower, he enveloped the front entrance to the entrenchment with fire and, after killing the occupants, ran through the smoldering tunnel to come out at another entrance. After the artillery piece had been destroyed with an explosive charge, he again entered the emplacement and fired his flame thrower into various openings within. Shortly thereafter, he manned a rocket launcher and destroyed an enemy tank as it approached our lines from behind the ridge.[35]
>
> — Navy Cross citation for PFC Robert Fransko

Sadly, clear evidence of the need for rifleman initiative now seems only a mundane "aside" for most 2GW-oriented historians. But, similar squad heroics had proven just as effective for 2/9.

> 27 February 1945. With his squad abruptly halted by intense fire from enemy fortifications in the high rocky ridges and crags commanding the line of advance, Private Watson [of 2/9] boldly rushed one pillbox and fired into the embrasure with his weapon, keeping the enemy pinned down single-handedly until he was in a position to hurl in a grenade and then running to the rear of the emplacement to destroy the retreating Japanese and enable his platoon to take its objective. Again pinned down at the foot of a small hill, he dauntlessly scaled the jagged incline under fierce mortar and machine-gun barrages and with his assistant automatic rifleman charged the crest of the hill, firing from his hip. Fighting furiously against Japanese troops attacking with grenades and knee-mortars from the reverse slope, he stood fearlessly erect in his exposed position to cover the hostile entrenchments and held the hill under savage fire for fifteen minutes, killing sixty Japanese before his ammunition was exhausted and his platoon was able to join him.[36]
>
> — MoH citation for Pvt. Wilson D. Watson

However, the battle for the foe's MLR was far from over. Even from a recently captured flank, that big an array of enemy strongpoints posed quite a threat. That's because each little element had its own all-around defense.

Occupying Airfield No. 2 at Center of Second Defense Belt

The hundreds of Leathernecks fighting for Airfield No. 2 soon realized—from the commanding terrain to its side and back—that the beating heart of the foe's main defensive effort was nearby.

> The runways of Motoyama Airfield No. 2 lay across the [3rd] Division zone of action, directly in front of the Ninth, providing perfect fields of fire for enemy machine guns and antitank guns on the flanks. . . . This position commanded the entire southern end of the island as far south as Mt. Suribachi, including Airfield No. 1.[37]
>
> —*Third Marine Division,* by Aurthur and Cohlmia

But, the east end of the airfield was in the 4th Division's zone of responsibility. Its big radio tower suggested the nearby presence of a command center. Luckily, both divisions had the perfect infantry leaders on hand to take the measure of the "belly of the beast." In the 4th Division zone, 2/24 had made the initial approach. Its commander had joined a small assault group to show all NCOs how important their job would be to final victory. As a former 1stSgt., he knew how powerful a Marine fireteam could be.

> 21 February 1945. . . . Capt. McCarthy [of G/2/24] acted on his own initiative when his company . . . was held up . . . during the approach to Motoyama Airfield No. 2. Quickly organizing a demolitions and flamethrower team to accompany his picked rifle squad, he . . . directed the combined operations of his small assault group, [until it] completely destroyed the hostile installation.[38]
>
> — MoH citation for former 1stSgt. and
> rifle company commander on Saipan
> Capt. Joseph J. McCarthy

Then, most of the fighting for the actual runways would be done by 3/21 and 3/9 from 24 to 26 February. (see Map 10.5.)

24 February 1945. Penetrating one of the most heavily fortified Japanese positions on Iwo Jima in an attempt to cut across Motoyama Airfield Number Two and capture the commanding ridge, Second Lieutenant Grossi [of K/3/21] and his men maneuvered . . . to the airstrip . . . and . . . surged across the heavily mined sands dotted with buried pillboxes. Determined to gain his objective . . . , Second Lieutenant Grossi led his unit against the Japanese with dauntless courage, slashing . . . with his bayonet, thrusting grenades into massive emplacements sunk in the sands, dropping them behind rocks, sweeping aside the enemy, holding desperately and fighting furiously in hand-to-hand engagements as he smashed through a sector swarming with Japanese to gain the fifty foot ridge on the opposite side of the airstrip after approximately ninety minutes of savage conflict. When a strong enemy counterattack forced him to yield the newly-won position, he promptly and effectively

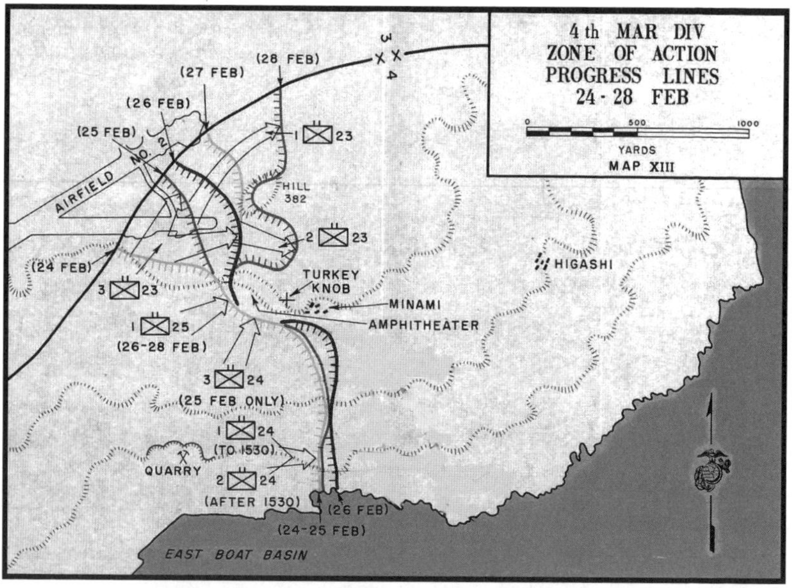

Map 10.5: The Taking of Airfield No. 2
(Source: Lt.Col. Whitman S. Bartley, "Iwo Jima: Amphibious Epic," Washington, D.C.: History Branch, HQMC, 1954, map XIII, p. 268.)

> reorganized the twelve men remaining in his unit, led the stouthearted group in a fierce bayonet and grenade charge and reoccupied the vital hill.[39]
>
> —Navy Cross citation for 2dLt.Dominick J. Grossi

> 25 February. . . . Gallantly risking his life under heavy enemy fire, Lieutenant Colonel Boehm [of 3/9] fearlessly carried out reconnaissance of hostile territory in preparation for attacks which resulted in the advance of his Battalion across Motoyama Airfield Number Two despite withering opposition and capture of a formidable Japanese position which had checked the advance of two divisions for several days.[40]
>
> —Navy Cross citation for Lt.Col. Harold C.Boehm

At that point, 4th Division once again got involved. The movement coordination between divisions had been hard enough, but now there was the heaviest preparatory fire so far on Iwo to worry about. The separate bombardments of two side-by-side U.S. divisions had put a lot of extra "flying metal" into the air. Upright travel and shrapnel don't mix well.

> The evening of February 25 . . . Company C [of 1/23] [was] dug in at the edge of the second airfield. The company line was just short of the runway and the unit leaders dug in at the bottom of the fill embankment which was probably about fifteen feet lower than the airport runway [itself]. . . . Sometime after midnight Homer [Booth] woke me [Rowland Lewis] . . . and whispered to me that there was a Japanese patrol at the top of the embankment. . . . I could see three Japanese clearly visible by the light of the flares. . . . At about that time, they were spotted and a flurry of fire that killed two of the three. . . .
>
> We jumped off [in the attack on the morning of the 26th] . . . with all three platoons abreast with the second platoon on the left, first platoon in the center and third platoon on the right. . . . [W]e double-timed across the airport runway since no one wanted to get pinned down on the runway where there would be no cover. As we crossed the runway, the [U.S.] artillery and naval gunfire lifted and started firing a couple of hundred yards ahead of us. As soon as we [the

> first platoon] were across the runway some of us saw the Japanese coming out of a large emplacement and presumably headed for their individual spider holes. They were taken under fire, but their exposure was so brief that no discernible damage was done before they got under cover and out of sight. My estimate is that the platoon suffered abut 20 killed or wound in the next few minutes. The slaughter stopped when we were able to get our attached machineguns in action. . . . The artillery and naval gunfire was called back in about 50 yards in front of us. Carrier planes came in one bombing and strafing runs maybe 50 or 60 yards in front of us. Unfortunately, none of his was of any direct help since our enemy was no more than 25 yards in front of us. . . .
>
> . . . Captain McDaniel told me [unit-leader Lewis] we were going to pull back across the airfield so they could call artillery fire into the area we were occupying. He told me to send someone over to tell the third platoon to pull back across the airport. . . .
>
> Richard Winegar was killed while trying to find the third platoon. . . . [W]hen we were pinned down, they continued to move which opened up a large gap between us.[41]
>
> — C/1/23 public domain website

After Airfield No. 2 had finally been occupied, there were hills and ridgelines just to the east that would require the Marines' full attention for several days. For a number of reasons (probably to include a few short-rounds), the whole area would soon become known as the "Meatgrinder."

Taking Charlie-Dog Ridge and Hill 382

Charlie Dog Ridge followed the southern edge of Airfield No. 2, whereas Hill 382 was just east of it. (Look back at Map 10.4) Guns embedded in the hill's crest had a perfect view of the airstrip.

> The top of the hill, surmounted by the stark remains of a Japanese radar station, was hollowed out and rebuilt to contain field pieces and antitank weapons. Each of these concrete gun housings was in turn protected by as many as ten supporting machine-gun emplacements. The rest

> of the hill was honeycombed with the same elaborate tunneling that characterized other major installations on the island. In addition, crevices and ridges crisscrossed the entire surrounding area. Light and medium tanks, armed with 57mm and 47mm guns, parked well back in these crevices, commanded the length of the main (northeast-southwest) runway of the airfield and approaches from the southwest.[42]
>
> — *Iwo Jima: Amphibious Epic*
> Historical Branch, HQMC

3/23 had been the first Marine battalion to approach Hill 382 from the southwest. In fact, one of its patrols had gone up it, and then come right back down after noticing a bypassed string of pillboxes.[43] On February 26, a tiny element from 3/23 did finally knock out those same fortifications. The element consisted of a lone rifleman—the aforementioned PFC Jacobson.

> February 26, 1945. Private First Class Jacobson [of 3/23] waged a relentless battle as his unit fought desperately toward the summit of Hill 382 in an effort to penetrate the heart of Japanese cross-island defenses. Employing his weapon with ready accuracy when his platoon was halted by overwhelming enemy fire on February 26, he first destroyed two hostile machine-gun positions, then attacked a large blockhouse, completely neutralizing the fortification before dispatching the five-man crew of a pillbox and exploding the installation with a terrific demolitions blast. Moving steadily forward, he wiped out an earth-covered rifle emplacement and, confronted by a cluster of similar emplacements which constituted the perimeter of enemy defenses in his assigned sector, fearlessly advanced, quickly reduced all six positions to a shambles, killed ten of the enemy and enabled our forces to occupy the strong point. Determined to widen the breach thus forced, he volunteered his services to an adjacent assault company, neutralized a pillbox holding up its advance, opened fire on a Japanese tank pouring a steady stream of bullets on one of our supporting tanks and smashed the enemy tank's gun turret in a brief but furious action culminating in a single-handed assault against still another blockhouse and the

> subsequent neutralization of its firepower. By his dauntless skill and valor, Private First Class Jacobson destroyed a total of sixteen enemy positions and annihilated approximately seventy-five Japanese.[44]
>
> — MoH citation for PFC Douglas T. Jacobson

Had PFC Jacobson been high an adrenaline, or just tired of seeing his buddies hurt through the 3-to-1 odds of a doctrinally correct assault? [45] Jacobson may have been just demonstrating how productive a smaller foray can be. But, even with Jacobson's help, the battle for Hill 382 was far from over for 3/23. Eventually, that battalion would have to withdraw for the night. And subsequent battlefield maps no longer show it as a participant.[46] A Japanese historian would later claim catastrophic damage to some Marine battalion, but which one was never made clear.[47]

> When his platoon leader [of I/3/23] became a casualty during a subsequent bitter action [on 26 February], he [PFC Bell] promptly assumed command and, after reorganizing his men, advanced with a companion to destroy a machine-gun nest which had been inflicting heavy casualties and retarding the advance of his platoon. Continuing his daring tactics when machine-gun fire emanating from a cave to the front of the line halted the advance, he again volunteered to rush the position and, hurling his grenades with deadly accuracy, assisted in killing three of the Japanese and in destroying a machine gun.[48]
>
> — Navy Cross citation for PFC Joshua S. Bell

1/23 had also been somewhat involved with 382 from the next lane over.

> February 26th. . . . Our battalion [1/23] was committed, and "C" company was to pass right through "K" Co. of the 3rd Battalion, and continue the attack on Hill #382. We moved out at 7:30. The Third Marine Division was to be on our left flank. They were not! Both [of] our flanks were exposed. . . . About 10:30 the enemy fire increased . . . causing heavy casualties. . . . Our heavy casualties . . . [were] caused by enemy positions firing . . . from Hill #382.[49]
>
> — C/1/23 public domain website

> February 27. . . . Early morning had "A" and "B" Companies taking over the attack and "C" Company . . . to be in reserve. However, we moved into an area on our left rear which had been bypassed by the 3rd Division. . . . We wound up killing 50 enemy men . . . and knocking out two machinegun nests.[50]
>
> — C/1/23 public domain website.

Hill 382 Had Been Part of MLR Cluster of Major Bastions

For a full week from 25 February to 2 March, 4th Division units would be largely stymied by all the mutually supporting fires from inside the Meatgrinder. The Marines attempting to take Hill 382 had been taking a lot of fire from elsewhere.

> 1 March 1945. Climbing to the top of a jagged, towering rock on Hill 382 during a fierce skirmish with the Japanese in one of the strongest enemy defensive sectors, Private Pound [of G/2/24] repeatedly exposed himself to intense rifle, grenade and machine-gun barrages in order to fire upon the many Japanese running intermittently from their caves in desperate delaying counterattacks against our advancing Marines. With hostile hand grenades exploding dangerously close to his vulnerable position and bullets whining past him, once piercing his helmet and twice damaging rifles in his hand, Private Pound maintained steady fire upon all Japanese coming within range of his deadly weapon and, making his way toward the rear area on two occasions for spare rifles and additional ammunition, returned each time to the rocky pinnacle and resumed his daring action, killing at lest two of the enemy and pinning many more inside their caves and emplacements. Taking up his station again, following a second trip to the rear lines, he continued to wage a bitter fight against the fanatic enemy, completing ninety minutes of almost constant exposure to intense fire before he fell, mortally wounded.[51]
>
> — Navy Cross citation for Private Henry T. Pound

Then, on 2 March, after six days of separately assaulting the Meatgrinder's key bastions (Hill 382, the Amphitheater, and Turkey Knob), the 4th Division commander decided to hit all three at once. (See Figure 10.1.) The main effort was to be by the 24th Marines against Hill 382, while the 23rd and 25th Marines assailed the Amphitheater and Turkey Knob. At 382, 2/24 would go up the back of the hill from the northwest, while 1/24 enveloped its front from the south. After a heavy fight, E and F companies finally reached the crest in early afternoon. But, as with the 23rd Marines' units before them, that didn't end the fighting. In fact, several more days would be necessary to consolidate the hill. That's because underground passageways had connected all pillboxes on its slopes. Every time one pillbox was damaged, more occupants would emerge from the hill's interior. While E and F Companies of 2/24 had mostly fought for 382, G also participated.[52]

At 0630 on 3 March, 4th Division resumed the attack all along the front without any preparatory bombardment. On the left, the 24th Marines faced a formidable defense on the high ground northeast of Hill 382. To 2/24's front lay reinforced concrete emplacements, one with a high-velocity gun. Down at the Amphitheater and Turkey Knob, 23rd Marines had just taken over the from the 25th. But, every avenue of approach had been covered by the foe's machineguns, and all defiladed areas by his mortars and artillery. Only with great difficulty could U.S. tanks move forward. Even then, they had limited fields of fire.[53] The belly of the beast would remain in tact a little while longer.

> 3 March 1945. [V]oluntarily undertaking the rescue of six wounded men of his platoon who lay exposed to devastating enemy fire, Assistant Cook Gambino of L/3/23] courageously pressed forward of the front lines through cross-fire from two heavy machine guns and, laboriously crawling and dragging each wounded man back to the protection of rocks, succeeded in rescuing five of the casualties. Despite a serious leg wound received on the fifth trip, he continued his heroic efforts until he lost his life in an attempt to rescue the sixth man.[54]
>
> — Navy Cross citation for Cpl. Frank J. Gambino

More About Turkey Knob and the Amphitheater

The same sort of desperate fighting—as at Hill 382—had been

going on at the Amphitheater and Turkey Knob. After 1/25 and 3/24 contested them from 26 February to 28 February, 1/25 and 1/23 did so from 1 March until 9 March. Wherever the Marines tried to breach an enemy strongpoint matrix, they paid for it dearly. Its occupants would pull back a little, and then call in preregistered indirect fires on where they had been. If the Americans were forced out by the approaching darkness, Japanese soldiers would sneak back into their former positions to prepare a nextday *encore*.[55]

Most of that initial effort had been against Turkey Knob. On 28 February, 25th Marines tried a pincers movement supported by two Sherman tanks. It failed, and the tanks' 75mm shells scarcely scratched the objective blockhouse. To 1/25's right, 2/25 had managed to haul in and reassemble a 75mm pack howitzer that could directly engage the concrete structure atop Turkey Knob. But then, over 40 more 75mm impacts at the same spot on its walls did little damage. That night, enemy artillery rounds (or sappers) scored several "lucky hits" on 5th Division ammunition dumps. Those big rear-area explosions did little to bolster frontline American morale.[56]

But, the Marines kept trying. Without any preparatory bombardment, 1/25 was to first "infiltrate" the high ground north of the tiny peak at 0630 on 2 March, and then elements of 1/25 and 3/23 (now attached to 25th Marines) would double-envelop it from the northwest and south. But, that unconventional attack was soon spotted and its participants forced to return to their starting points (some under cover of a smoke screen). While Marine artillery and mortars retaliated, eight Sherman tanks moved in to pummel the Turkey Knob blockhouse. Any number of 75mm shells and 1,000 gallons of flamethrower fuel were hurled against it with no lasting effect. The emplacement was being remanned from the hill's interior. Finally, at 0630 on 3 March, 1/23 passed through 1/25 and again attacked without any preliminary bombardment. And after a little artillery support, it re-assaulted. Both times, it ran into too much fire from the blockhouse.[57]

> 3 March 1945. When his company was ordered to seize a ridge to which the enemy had clung tenaciously and from which devastating fire had prevented two previous attempts to dislodge them, Private First Class Hahn [of A/1/23] realized that . . . it would be impossible to deliver

> effective machine-gun fire on the enemy from an emplaced position. Standing upright, he began firing from the hip, catching the enemy by his sudden maneuver and forcing them to break from their positions and commence withdrawing. Although other men were wounded and killed within a few yards of him, Private First Class Hahn never wavered in his determination to carry out his mission and, expending seven hundred rounds of ammunition, inflicted at least twenty-five casualties upon the enemy and effectively neutralized this pocket of resistance, thereby enabling his company to seize the ridge.[58]
>
> — Navy Cross citation for PFC Orville W. Hahn

The foe had now mined all approaches to Turkey Knob. Any attempt to remove those antipersonnel devices was met by deadly sniper fire. But, by 1400 on 3 March, a U.S. engineer platoon had cleared a way for flame tanks and demolition teams to get within range of the fortification. Then, repeated infantry assaults—over the course of the afternoon—did partially reduce the objective. But when the grunts were ordered to realign for the night, Turkey Knob came again under Japanese control.[59]

On 6 March, the American attack resumed—after an earth-shaking naval-gunfire and artillery barrage—with 1/25 again in the lead. Once more it failed. Turkey Knob and the Amphitheater would have to be temporarily bypassed for less costly gains elsewhere. Not until 9 March, would the former become sufficiently surrounded by U.S. units to permit its systematic reduction.[60] Of course, having many of its defenders accompany Gen. Senda to the east greatly helped.

Map 10.1 may have identified Mount Byobu as a tiny Charlie-Dog Ridge promontory next to the airport. But, it was more likely Turkey Knob, some 700 yards to the east. Here's how a Japanese historian later described the Mount.

> In the battle for Mount Byobu, hand-to-hand combat continued for three days and three nights, during which the American forces seized the hill six times, and the Japanese reclaimed it five.[61]
>
> — *From Uraga to San Francisco,* by Okazaki

Turkey Knob had been Senda's headquarters for the Second

Figure 10.1: Turkey Knob
(Source: Alexander, "Closing In: Marines in the Seizure of Iwo Jima," Washington, D.C.: History and Museums Division, HQMC, 1994, p. 35.)

Mixed Brigade.[62] Right before it fell around 9 March, he and many soldiers exited the open-ended cordon around both it and the Amphitheater.

> One exception [to Kuribayashi's *banzai* ban] occurred the night of 8 March when General Senda grew so frustrated at the tightening noose being applied by the 4th Marine Division that he led 800 of his surviving troops in a ferocious counterattack.[63]
>
> — *Closing in: Marines in the Seizure of Iwo Jima*
> History and Museums Division, HQMC

After being last assailed on 3 March, most the Amphitheater remained in Japanese hands until around 9 March.[64]

Place to Where Gen. Senda Had Moved His Headquarters

The eastern portion of the Second Belt was to be contested for another week. Its trace would be defined by where Senda moved his headquarters. (See Map 10.6 and then look back at Map 10.4.) The Turkey Knob breakout had been above ground and quickly countered.[65] So, its participants would have needed a tunnel to travel the remaining 1000 yards without mishap.

The fighting for Senda's new pocket would start around 10 March, south of Higashi. It turned out to be not only a full-fledged defense matrix with sub-surface honeycomb,[66] but also fairly well stocked with supplies.

> When his [PFC Ioanna's] platoon [of F/2/25] was held up by heavy Japanese fire on 10 March 1945, he voluntarily joined a bazooka operator in crawling forward to neutralize the hostile pillbox and, after his comrade was killed, picked up the bazooka with which he had no previous experience, moved to within fifty feet of the enemy and destroyed the hostile strong point. On 12 March, he participated in a scouting mission behind enemy lines and succeeded in destroying a large Japanese ammunition dump.[67]
>
> — Navy Cross Citation for PFC John G. Ioanna

The landscape at Senda's new location would be much like where Kuribayashi ended up on the north coast.

> The enemy-held terrain consisted of a series of deep crevices and sharp ridges running in a generally southeasterly direction toward the coast. Other lesser gullies cut transversely through the area to create a maze of compartments and cross-compartments. Rough rocky outcroppings and scrubby vegetation . . . provided excellent cover and concealment to the defenders (4thMarDiv D-2 Periodic Rpt. No. 69, 12 Mar 45).[68]
>
> — *Iwo Jima: The Amphibious Epic*
> Historical Branch, HQMC

Senda's final location—the southern equivalent of Kuribayashi's Gorge—would not be reduced until 16 March. Its preparations had been so complete (to include all the ordnance) that it must have been an original bastion in the Nipponese Second Defense Belt.

The plan of attack called for 2/25 to attack down the draws toward the coast while 3/25 and 2/24 supported the attack by firing bazookas, anti-tank grenades, and 60mm mortars into the area to their front. Following this general scheme, the regiment slugged it out with the Japanese in the pocket for four days. Tanks operated with caution in supporting the attack . . . but both regular and flame throwing mediums contributed in large measure to the eventual success of the infantry. . . . As the operation against the pocket continued, the main core of the resistance developed northwest of the "beach road." . . . Plans were made to send tanks out along the road to positions in [the] rear of the pocket to fire north into the draws in an attempt to knock out hostile mortar positions.

On 15 March, Colonel Lanigan, commanding RCT 25, shifted the main pressure against the pocket to the south and directed the skeleton companies of the two battalions in the center of the line to attack regardless of contact. This drive gained 200 yards and penetrated deeply into the left flank of the enemy's strongest positions, paving the way for elimination of the stronghold the next day.

During the night enemy troops made several attempts to infiltrate Marine lines from the south, but each time the movement was detected and broken up. When the 25th jumped off at 0630 on 16 March, the Japanese fought back . . . , but now, without any real organization left, they resisted only in small isolated groups. By 1030 all battalions had cut through to the beach road, and RCT 25 announced complete destruction of all resistance in the pocket.

Although the original location of General Senda's Brigade Headquarters was just east of Hill 382, there is little doubt that his final command post was in this area where some 1,500 Japanese troops . . . defended to the end.[69]

— *Iwo Jima: The Amphibious Epic*
Historical Branch, HQMC

As Senda's pocket collapsed, a U.S. corpsman further confirmed that its caves had been at the center of an MLR bastion.

[O]n 15 and 16 March 1945. Almost continuously under fire

> while carrying out the most dangerous volunteer assignments, Pharmacist's Mate First Class Pierce [of 2/24 had] gained valuable knowledge of the terrain and disposition of troops. Caught in heavy enemy rifle and machinegun fire . . . , Pierce deliberately exposed himself to draw the attacker from the cave and destroyed him with the last of his ammunition. Then lifting the wounded man [he had been trying to save] to his back, he advanced unarmed through deadly rifle fire across 200 feet of open terrain. Despite exhaustion . . . , he again traversed the same fire-swept path to rescue the remaining Marine. On the following morning, he led a combat patrol to the sniper nest [outlying position of a strongpoint] and, while aiding a stricken Marine, was seriously wounded.[70]
>
> —MoH citation for Pharmacist's Mate First Class Francis Pierce, Jr.

Hard-to-Explain Actions of Top Japanese Commanders

Why did Senda move his CP from the middle of the Second Defense Belt to its southeastern end? And, why had Kuribayashi just shifted his from behind where Senda had been to the north end of the Third Belt? While facing the foe's main effort, one had gone left and the other right. Neither had wanted to survive the battle. Why would they both abandon their men at such an inauspicious time? Might they have been separately drawing attention away from the island-long thoroughfare? Where that oversized tunnel led beyond the Third Belt and its possible utility to wounded Japanese soldiers would never be investigated. That's largely because no top enemy commander ever made a radio transmission from there.

Both Japanese generals had been well-versed in battlefield deception. Dummy positions to absorb U.S. shelling and bombing had admittedly been part of their plan.[71]

Before moving, Gen. Senda had been "just east of hill 382" at Turkey Knob,[72] and Gen. Kuribayashi somewhere between Senda and Airfield No. 3. The History and Museums Division at HQMC says that before occupying the Gorge cave, Kuribayashi had been near Motoyama Village.[73]

> On this date [4 March], Kuribayashi [had] moved his own command post from the central highlands to a large cave on the northwest coast.[74]
>
> — *Closing in: Marines in the Seizure of Iwo Jima*
> History and Museums Division, HQMC

That's where Lt.Gen. Kuribayashi joined Col. Ikeda (CO of 145th Infantry Regiment) and Admiral Ichimaru (overall commander of the Naval units on Iwo) in the same CP near Kitano Point.[75] All were familiar with the teachings of Sun Tzu, so they knew that where opposition forces were united, those forces would have to be separated.[76] They also knew that too lengthy a U.S. presence above a hidden evacuation route would eventually lead to its discovery.

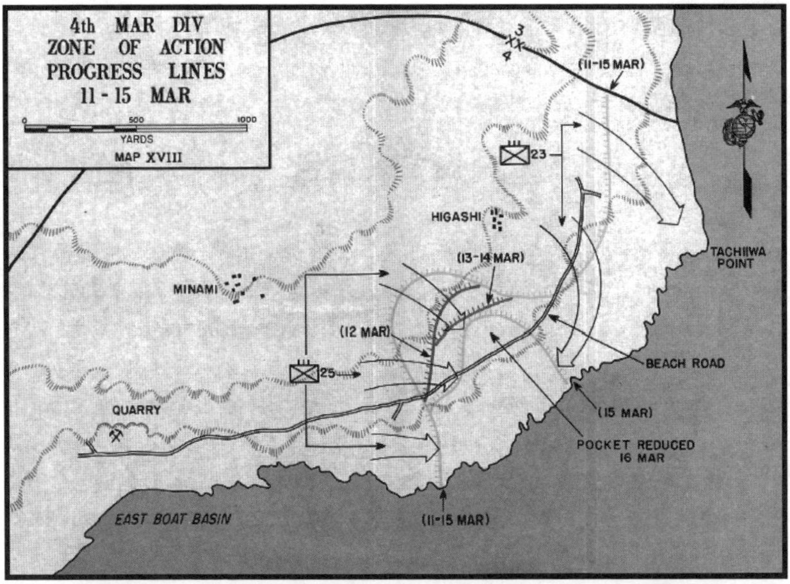

Map 10.6: Final Pocket to Which Senda Moved His CP
(Source: Lt.Col. Whitman S. Bartley, "Iwo Jima: Amphibious Epic," Washington, D.C.: History Branch, HQMC, 1954, map XVIII, p. 272.)

11 WITH REGARD TO THE THIRD DEFENSE BELT AND GORGE

- Did the Third Defense Belt exist, or just an oblong zone?
- What was at either end of that Belt?

The Gorge defenders' fire plan was masterful.

(Source: Charles Waterhouse painting of the Gorge, as part of "Closing In," by Col. Alexander, Marines in WWII Commemorative Series, HQMC pamphlet, 1994.)

What Lay Beyond Airfield No. 2

As the Marine divisions moved beyond the foe's MLR, their most experienced frontline troops had already become casualties. But, the hastily trained replacements were eager to try new things. So, the unconventional use of a few infiltrators to silence enemy machineguns continued for many assault units. While 4th Division was still busy with Senda's last stand in the Second Belt, the other two divisions headed for their assigned sectors in the presumed Third. (Refer back to Map 10.2.) Map 1.1 had depicted that Third running from 1000 yards north of Nishi Village to just south of Tachiiwa Point. An official battle chronicle would provide a slightly more detailed path. "It began several hundred yards below Kitano Point on the northwest coast, cut through Airfield No. 3 and the Motoyama [Village] area in the center, and terminated between Tachiiwa Point and the East Boat Basin on the eastern coast."[1] More specifically, it ran from the Gorge, to Hills 362-B, 357, 331 362-C, and then somewhere below Tachiiwa Point. Within that central cluster of fortified hills would eventually lie Cushman's Pocket.

> The 5th Division would . . . lose most of its invaluable [dwindling] cadre of veteran leaders attacking . . . 362-B, then end the fighting in "The Gorge." The 3d Division would focus first on . . . Hill 362-C . . . , and finally the moonscape jungle of stone which would become known as "Cushman's Pocket."[2]
>
> — *Closing In: Marines in the Seizure of Iwo Jima*
> History and Museums Division, HQMC

Where the Third Defense Belt Went After Airfield No. 3

The post-war Japanese map claimed no third cross-island defense line at all, but only an oval 2000-yard-wide defense zone called the "Honeycomb Defense Position."[3] (Look back at Map 10.1.) Such a zone was certainly possible, because a fortified ring had been expected at the front of Suribachi. (Look back at Map 1.1.) And the last major resistance on Iwo would occur inside that Honeycomb.

Yet, big fights had also happened outside the purported oval. Cushman's Pocket was to hold out until 16 March. And the 2nd Mixed Brigade intercept in Chapter 10 had discussed defending Iwo all the way to its southeastern coast. Here's how another U.S. chronicle described the fortification.

> The . . . [third] defense line generally bisected the remaining area in the northern portion of the island. . . . Man-made emplacements were not as numerous in this . . . line, but natural caves and other covered positions afforded by the fantastically rugged terrain were skillfully organized

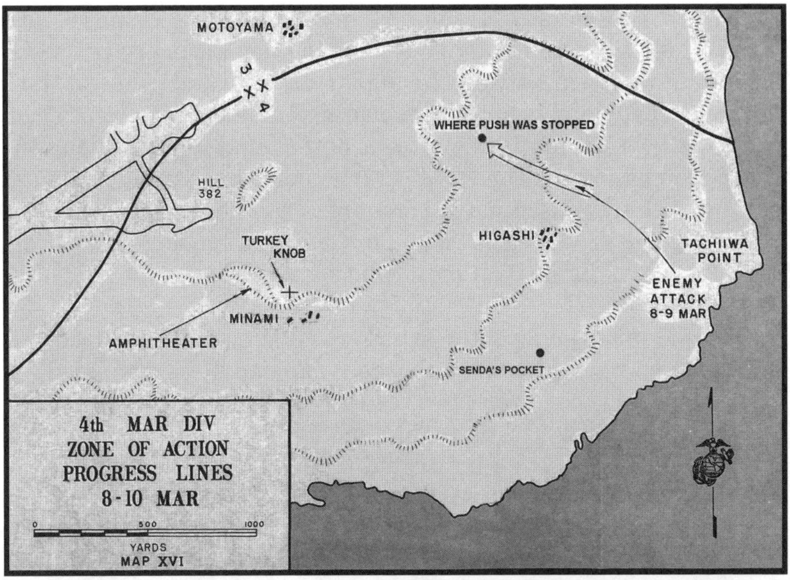

Map 11.1: Massive Japanese Counterattack of 8-9 March
(Source: Lt.Col. Whitman S. Bartley, "Iwo Jima: Amphibious Epic," Washington, D.C.: History Branch, HQMC, 1954, map XVI.)

for the defense.[4]

— *Iwo Jima: Amphibious Epic*
Historical Branch, HQMC

Other engagements closer to Tachiiwa Point would help to define the Japanese "line's" final path.

Evidence of Another Third-Belt Bastion

There had been a massive counterattack by Capt. Samaji Inouye's Naval Guard Force some 1000 yards north of Senda's last stand (on the Second Belt) during the night of 8-9 March. (See Map 11.1.) That this attack had been headed northwest belies the purported purpose of destroying airplanes or recapturing Suribachi. The big enemy push was finally stopped very near to where Map

2.2 had shown a Navy Headquarters on the Third Defense Line. (See Maps 11.2 and 11.3.) That's too big a coincidence not to indicate a connection between headquarters role and counterattack purpose. Instead of honorably expiring, that 800-1000 man contingent had been trying to reinforce a parent-unit bastion along the Third Defense Belt. There was only a small infantry detachment with those hundreds of naval personnel.[5] And to further delay the U.S. invasion of the home islands, Gen. Kuribayashi had—from the very start—banned all mass suicides.

> Although some of the attackers shouted and screamed, this was no wild *banzai* charge. The Japanese had used the twisted, shell-torn terrain to good advantage as they worked their way close to the lines before commencing the assault. A few even infiltrated through the front lines to the 2/23 command post where they harassed the operations section with hand grenades. . . . Close-quarter fighting raged throughout the hours of darkness with Company E bearing the brunt of the attack, but heavy fighting also took place in Company L, 3/24's sector and elsewhere, along the line.
>
> By daylight the struggle had nearly died out, but mopping up continued until noon. Approximately 650 enemy dead were counted in the area of the main attack, and reports from other [nearby] sectors brought the total to nearly 800. . . . Maps and documents recovered from the dead revealed the major Japanese units in the counterattack to be the 1st Company, 310th Independent Infantry Battalion [with 150 men at best] . . . , and [the rest mostly] naval units. . . .
>
> . . . It is perfectly plausible that these troops . . . hoped to join [other] forces in the area.[6]
>
> — *Iwo Jima: Amphibious Epic*
> Historical Branch, HQMC

Last Bastion on That Third Belt

On 10 March, 2/23 ran into big trouble near Tachiiwa Point. Because of the kind of terrain and resistance, this location may have anchored the east end of the final defense line.

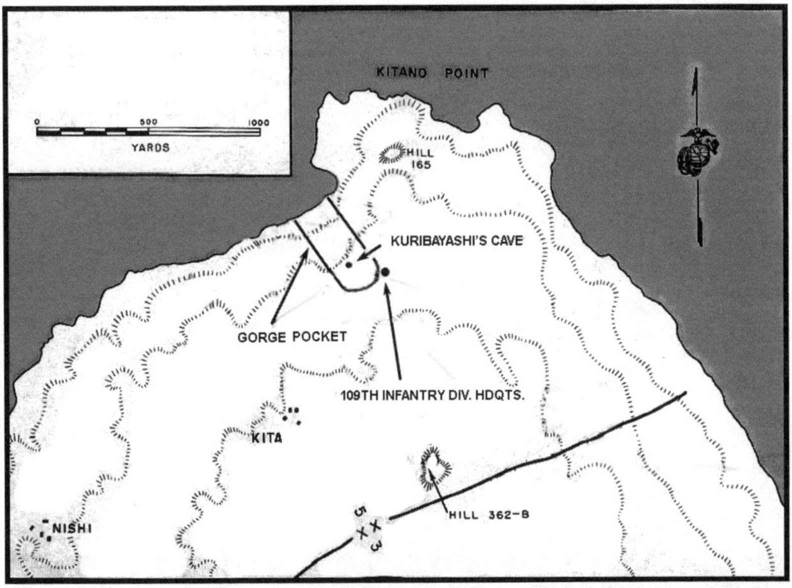

Map 11.2: Terrain and Events at Top of Third Defense Belt
(Source: Lt.Col. Whitman S. Bartley, "Iwo Jima: Amphibious Epic," Washington, D.C.: History Branch, HQMC, 1954, map XII.)

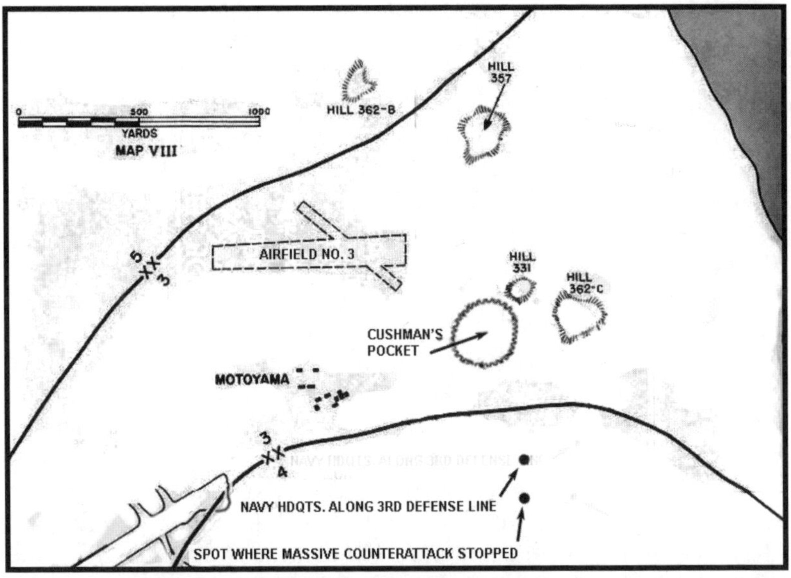

Map 11.3: Terrain and Events at Bottom of Third Defense Belt
(Source: Lt.Col. Whitman S. Bartley, "Iwo Jima: Amphibious Epic," Washington, D.C.: History Branch, HQMC, 1954, map VIII.)

> 10 March 1945. When intense enemy machine-gun fire held up the advance of his platoon, Private First Class Latch [of G/2/23] located a strongly fortified cliff-line position and, although other attempts to destroy the emplacement failed, charged twice, unarmed, in the face of fierce hostile machine-gun and sniper fire to place demolition charges at the mouth of the cave. As these charges failed to silence the Japanese, he scaled the cliff to a position above the hostile cave and twice lowered himself at great risk to drop two more demolition charges with the result that he destroyed the position and the enemy within.[7]
>
> — Navy Cross citation for PFC Daris W. Latch

What Role Did Kita Village Play in the Third Belt?

Chapter 2 has made mention of a defense matrix at Kita Village near Iwo's north coast. That matrix could have been on the western edge of a defense zone. It could have been part of a cross-island strongpoint belt. Or, it could have just been a dummy facility to draw American fire away from key sites. Whatever its purpose, it caused TF 56 planners (the makers of Map 1.1) to believe the Third Defense Line began well south of the yet-to-be-famous Gorge.

That Kita Village complex had been cursorily searched (probably by a Marine unit *en route* to the Gorge). It sported a large underground chamber with "500-foot exit tunnel that led to an opening between small hills." With 20 radios inside, this chamber was—by all outward appearances—not only a Naval Land Force CP, but also the hub for all communications with IJA Headquarters in Japan. (Look back at Map 2.2.) Yet, "[t]he alternate command post for the Naval Land Force and [primary CP for] 109th Division . . . [had been] dug . . . about 500 yards northeast of Kita." That puts them inside the Gorge and near the cement dome atop CR 291. That Kita Village tunnel had likely headed in the same direction as those other CPs. (See Map 11.4.) After being initially sealed, Kuribayashi's cave would be secretly remanned (more on that later).[8] So, this Kita Village tunnel had major significance. After one end had fallen behind U.S. lines, the other end could have greatly facilitated any Gorge breakout.

Map 11.4: The Gorge in Relation to Hill 291 (Table Rock)
(Source: G-2 FMFPac "Report of Intelligence," map from this url: http://www.ibiblio.org/hyperwar/PTO/Iwo/ComTaskForFiftySixIwoJima/TaskForFiftySix-EnclosureC.pdf.)

There Were Northern Ridgelines for Defense Belt to Follow

The northern end of that Third Belt probably ran along the rocky ridges that are known to have led into the Gorge's south end from the island's center. As with the Second Belt, such ridgelines probably formed the backbone for a Third.

> The 28th Marines [of the Fifth Division] inched forward . . . up the gradual slope toward the northwest-southeast ridgeline overlooking the [G]orge to its front. Atop this ridge the Japanese held strong fortifications that commanded all approaches from the southwest, and when Marine bombardment became intolerable they took refuge in caves that honeycombed the steep sides of the [G]orge. . . .
>
> . . . The most practical direction of attack was now from east to west paralleling the ridgelines that ran from the center of the island to the coast.[9]
>
> — *Iwo Jima: Amphibious Epic*
> Historical Branch, HQMC

The battles for key parts of this Third Belt will again be discussed from north to south to help establish its footprint. Because all three divisions were involved, those battles did not happen in that same order. In fact, the top of the Belt would be captured last.

Gorge and Its Ridgeline At the Very Top of Third Belt

Heavy fighting for the coastline end of the Gorge would not begin until 0830 on 11 March. (See Map 11.5.) Following a 50-minute preparation by air, naval gunfire, and 12 artillery battalions, seven 5th Division battalions attacked. On the left, those from the 28th Marines headed towards the ridgeline overlooking a large draw running east and west across the entire front. But, they encountered so much small-arms fire that only a 20-yard advance was possible. Another assault at 1300 netted 10 yards more. The overall push resumed the next morning with 27th Marines on the right. Again, it ran into too much enemy fire to make any progress. Another heavily supported attempt at 0900 was to no avail.

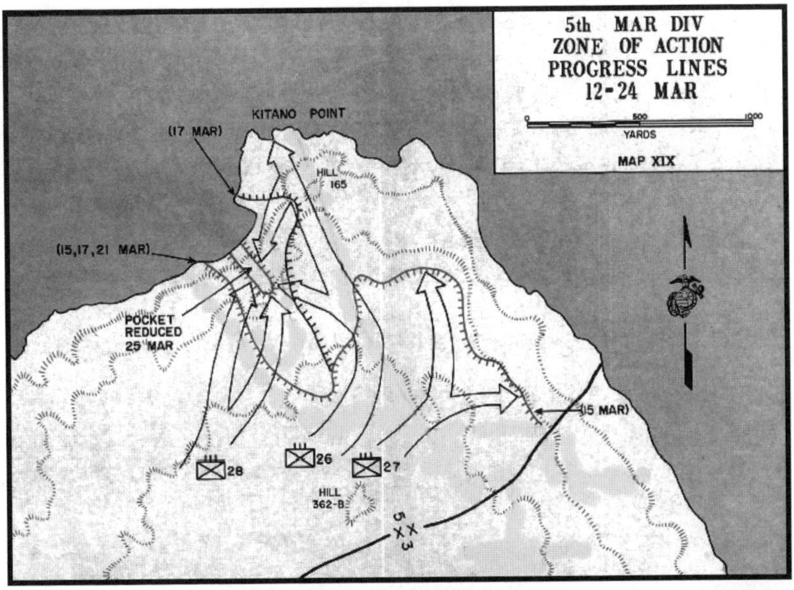

Map 11.5: Gorge Reduction
(Source: Lt.Col. Whitman S. Bartley, "Iwo Jima: Amphibious Epic," Washington, D.C.: History Branch, HQMC, 1954, map XIX.)

That's when 2/28—like at Suribachi—decided to send a patrol into the large draw at its right front. That patrol soon reported: "Numerous caves in the cliff line [when] facing north, of the ridge we were trying to reach." (This undoubtedly meant gun sites along the north rim of what would soon become known as the Gorge, because "caves are known to have honeycombed its steep sides.") The overall formation moved forward. After a further advance of 75 yards on the right, C Company of 5th Tank Battalion was able to neutralize a few pillboxes with flame by 1630.

The top of that objective ridge had stretched from the shoreline to hill mass CR (Collection Request) 291. A 5th Division website would later acknowledge the following. "The south [end] slope of the ridgeline was fairly gradual, but cut with many small draws parallel to it. The top was heavily fortified with pillboxes and spider traps, and the north [end] side was a steep cliff 40 feet to 50 feet high, honeycombed with many caves. It is believed that this ridgeline was the western sector of the enemy's final defensive line on northern Iwo."[10] From that excerpt's context, one can see that only the western side of that big ridge was being described.

What Made the Gorge So Difficult to Take

The fight for the Gorge would drag on until 26 March. That's because something unusual had happened. Individual defenders had been over-achieving. The close confines of the Gorge had provided them with a showcase for careful planning and advanced technique. Among other things, they had developed a fire plan for incrementally moving backwards. But, no reference would later claim their initial number to have been more than 600.[11] That's far too few for the island-wide last stand of a 22,000-man army. Yet, enough light-infantry skill could have created that illusion.

> In attacking these positions, no Japanese were to be seen, all being in caves or crevices in the rocks and so disposed as to give an all-around interlocking, ghost-like defense to each small compartment. Attacking troops were subjected to fire from flanks and rear more than from their front. It was always difficult and often impossible to locate exactly where defensive fires originated. The field of fire of the individual Japanese defender in his cave position was often limited to an arc of 10° or less, conversely he was protected from fire except that coming back on this arc. The Japanese smokeless, flashless powder for small arms, always advantageous, was of particular usefulness here. When the position was overrun or threatened, the enemy retreated further into his caves where he usually was safe from gunfire, only to pop out again as soon as the occasion warranted unless the cave was immediately blown (5th MarDiv IntelRpt,27).[12]
>
> — *Iwo Jima: Amphibious Epic*
> Historical Branch, HQMC

Here's how the dangerous area had looked to members of the 28th Marine Regiment as they began to shrink the enemy holdings.

8 March 1945. When his platoon was halted by heavy machine-gun and rifle fire which wounded the Platoon Leader and all Senior Non-Commissioned Officers, Corporal Mueller [of B/1/28] promptly and skillfully reorganized the platoon and, braving intense hostile fire, crawled from one foxhole to another to encourage his men to advance. Wounded while leading the platoon in an assault on an enemy-held ridge which was interwoven with tunnels and connecting trenches, he refused evacuation until the Japanese were routed from the position and thirty of the enemy were killed.[13]

—Navy Cross citation for Sgt.Bernhardt L.Mueller

9 March 1945. Constantly exposed to intense enemy hand grenade fire, Sergeant Kost [of C/1/28] advanced rapidly across open terrain to assault a hostile pillbox from which a barrage of hand grenades was falling on his organization. Upon reaching the enemy position, he dropped hand grenades into the embrasures of the pillbox, annihilating six of the Japanese . . . although severely wounded during the action.[14]

– Navy Cross citation for Sgt. Michael R. Kost

11 March 1945. When the advance of his company was stopped by withering cross-fire from enemy machine guns emplaced in pillboxes, Private Bleau [of D/2/28] voluntarily went to the rear and brought up an antitank rocket launcher. Starting in the sector on the right flank of his company, he attacked and destroyed two hostile pillboxes. Although constantly exposed to heavy enemy machine-gun fire, he continued across his company's front and destroyed three more similar Japanese emplacements. He was seriously wounded while attempting to close in on another hostile pillbox.[15]

– Navy Cross citation for Pvt. Calvin J. Bleau

14 March 1945. When the front line assault troops were pinned down by intense enemy mortar, machine-gun and rifle fire, Private Tankersley [acting squad leader of H/3/28] voluntarily left the company command post, obtained a flame thrower from the battalion dump and, on his own initiative, worked his way ahead of the front lines to attack entrenched

> hostile troops, destroying some of their positions and forcing them to abandon others.[16]
>
> — Navy Cross citation for Pvt. Oral L. Tankersley

Action Along the Backside of the Gorge Ridgeline

Southeast of the Gorge, 26th Marines had already severed the defense line and was attacking northward along it. (Look back at Map 11.5.) That again would take U.S. rifleman initiative.

> 14 March 1945. When heavy and accurate machine-gun and sniper fire held up the advance of his company and inflicted severe casualties in his platoon, Private First Class Wilson [of I/3/26] volunteered to act as guide for the tanks after his Company Commander requested tank support. Exposing himself to hostile fire near the tanks to use telephone communication with the Tank Commander, he crept and crawled over fire-swept open terrain for fifty yards to reconnoiter and to guide the first tank into an advantageous firing position.[17]
>
> — Navy Cross citation for PFC Ray E. Wilson, Jr.

> 15 March 1945. When the enemy could not be dislodged by rifle fire or grenades during an attempt by our troops to silence a heavy gun, Private First Class Harrison [of H/3/26] led his group in a charge on the hostile position and, although flash-blinded by an enemy grenade and unable to see clearly, braved a constant hail of rifle fire and grenades to engage in hand-to-hand combat with the Japanese. Personally annihilating five of the enemy, he forced the others to take cover in a deep cave and, leaving his men to guard the cave, ran back through hostile rifle and machine-gun fire to obtain explosives. Carrying them forward to the cave, he placed a charge to destroy the installation and, although wounded by rifle fire, remained in his exposed position until the charge was set.[18]
>
> — Navy Cross citation for PFC Henry A. Harrison

Before long, 2/26 found itself approaching the Gorge's south end. The well-fortified hill-mass CR 291 stood directly in its path.

> 13 to 16 March 1945. When all company officers had been rendered casualties in a drive on formidable and well-concealed enemy positions on 13 March, Sergeant Faltyn [of D/2/26] unhesitatingly assumed command of the Company. Fearlessly running from platoon to platoon to give orders and encourage his men in the face of intense hostile fire from numerous caves on high ground to the front, he reorganized the company and boldly led it forward to seize the objective. Although wounded in a fierce hand grenade exchange the next day, he directed the building of a road leading to the Japanese positions, organized tanks for support, and personally led his men in a long advance against a series of hostile caves and pillboxes.[19]
>
> — Navy Cross citation for Sgt. Hubert J. Faltyn

> 14 March 1945. Voluntarily taking command of his rifle squad when the leader became a casualty, Private Sigler [of 2/26] fearlessly led a bold charge against an enemy gun installation which had held up the advance of his company for several days and, reaching the position in advance of the others, assailed the emplacement with hand grenades and personally annihilated the entire crew. As additional Japanese troops opened fire from concealed tunnels and caves above, he quickly scaled the rocks leading to the attacking guns, surprised the enemy with a furious one-man assault.[20]
>
> — MoH citation for Pvt. Franklin Sigler

> 15 March 1945. Volunteering for duty as a Rifleman after his company suffered severe casualties, Private Palmer [of D/2/26] advanced with his platoon and suddenly came under heavy fire from the enemy concealed in spider traps and pillboxes. Quickly exhausting his supply of grenades in the ensuing firefight, he gathered other grenades from members of the platoon and, boldly charging the hostile positions, annihilated or routed many of the Japanese.[21]
>
> — Navy Cross citation for Pvt. Frank L. Palmer

With tank and dozer support, the right end of the 26th Marines formation moved northward on 17 March—planning to exploit any "soft spots." It got all the way to the coast near Kitano Point and then pivoted to the west. Strong enemy resistance had held up the formation's left end in a deeply ravined pocket of roughly 800x200 yards in size. No artillery or rockets could be fired on this small a target without risking friendly lives.[22] It would be up to frontline, low-ranking ground troops to reduce the pocket.

> 17 to 26 March 1945. Assuming the duties of a platoon leader when the leaders of two platoons had become casualties during a fierce grenade and small-arms battle with the enemy, Private First Class Little [of H/3/26] risked his life to move across the front of two platoons in the face of exploding grenades and small-arms fire to organize the remaining men into one unit.[23]
>
> — Navy Cross citation for PFC James T. Little

Finally, 1/28 and 1/26 were able to squeeze pocket defenders into a small enough space around 26 March to overwhelm whoever was left after a previous-night exfiltration.[24]

Had Hill 165 Been Part of the Belt Top?

Hill 165 lay slightly to the northeast of the infamous Gorge, but was not part of the Third Defense Belt. It had been too easily bypassed by the 26th Marines.

> With his battalion reduced to forty-three percent of its landing strength . . . during the fifth consecutive day of combat against strong enemy defenses concealed by camouflage in extremely rough terrain, Lieutenant Colonel Pollock (of 1/26) moved forward into the front lines under intense fire and, conducting a thorough personal reconnaissance of this dangerous area, prepared a bold enveloping maneuver [of the Gorge] aimed [initially] toward positions behind [just northwest of] Hill 165.[25] (Look aback at Map 11.5.)
>
> —Navy Cross citation for Lt.Col. Daniel C. Pollock

Fight for the Rest of the Southeasterly Running Ridgeline

The Japanese defense belt between the Gorge and Hill 362-B

followed a series of ridgelines, none of which were easy to occupy. Once again, low-ranking volunteers would help to make that possible.

> [W]hen his [PFC Coleman's' company [A/1/26] was again pinned down by the enemy's smashing barrage of fire on 7 March, [he started] cautiously infiltrating the Japanese lines to reach an extremely dangerous position in the midst of hostile emplacements located on the high ground behind the enemy's defensive position. Instantly attacking with hand grenades, he knocked out enemy guns, killed several of the Japanese and completely demoralized the hostile.[26]
>
> —Navy Cross citation for PFC Edward J. Coleman

Action upon Hill 362-B

Hill 362-B would be captured by 2/26 and 2/27 around 3 March. Initially, 3/9 had crossed the divisional boundary to occupy the hill's lower slopes. Then, it was replaced by 2/26. Before long, 26th Marines' right flank extended all the way across a very rough and craggy hilltop. Close-quarters fighting ensued on that broken crest. A company from 2/26 spent the night of 3-4 March throwing hand grenades at opposition soldiers only a few yards away. In fact, during the rest of the struggle for Iwo, grenade exchanges would be more common, because the island's terrain had become so broken. The greater number of caves at its north end had slowed the advance and made more demolitions necessary.[27]

> [O]n 3 March, he [Capt. Caldwell] [of F/2/26] led his company in a six hundred yard advance under heavy fire which inflicted extremely high casualties among his men. When his platoon leaders became casualties and the platoons became disorganized, he personally organized and maintained contact in his units despite hostile fire, and with unflagging determination and courage, inspired his men to hold the objective.[28]
>
> —Navy Cross citation for Capt. Frank C. Caldwell

> [O]n 3 March 1945. During an assault on a strongly held hostile position, Second Lieutenant Gelshenen [of F/2/26]

> led his platoon across an open field under intense Japanese artillery and small-arms fire and, reaching the base of the enemy-held cliff, quickly reorganized his platoon and commenced the ascent of the cliff under heavy enemy opposition. As he led his platoon up through numerous ravines, the Japanese fired an intense barrage of various caliber weapons, inflicting heavy casualties on the platoon. Undeterred by the savage opposition, he reorganized his depleted force for continuing the advance to the summit, and while so doing was severely wounded by shrapnel.[29]
>
> — Navy Cross citation for 2ndLt. M.L. Gelshenen

> Wounded again on 2 March, he [Corpsman Gehlen of F/2/26] . . . gallantly refused evacuation, moving out with his company the following day [3 March] in a furious assault across 600 yards of open terrain and repeatedly rendering medical aid while exposed to the blasting fury of powerful Japanese guns. . . [He persevered in his determined efforts as his unit waged fierce battle and, unable to walk after sustaining a third agonizing wound, resolutely crawled 50 yards to administer first aid to still another fallen fighter.[30]
>
> — MoH for Pharmacist's Mate Second Class George Edward Gehlen

Gen. Erskine had correctly guessed that the foe's Third Defense Belt would be less strong at Hill 362-B than closer to Airfield No. 3. So, he planned to move from just south of 362-B to Hill 357, and then on to 362-C from the flank. (Look back at Map 11.3).

The Fight for Hill 357 Behind Airfield No. 3

2/21 would capture Hill 357 by 1500 on 3 March and then unsuccessfully try to move toward Hill 362-C. There were no officially acknowledged acts of bravery, but a battlefield chronicle would indicate Hill 357 was fortified.

> The left [3rd Marine Division] regiment jumped off . . . with 2/21 in the attack. . . . As Companies F and G advanced, they received high-velocity fire from Hill 362-B, but kept

> moving ahead. By 0900 they had gone nearly 400 yards, and were approaching . . . Hill 357 at the eastern edge of the Motoyama Plateau. To cover the flanks of this attack, Company I, 3/9 [attached to 2/21] moved up on the left, and Company E (2/21 reserve) came up along the other side. At 1000, the 2d Battalion gained the top of the hill and began mopping up with the help of flame throwers and demolition teams from Company I.[31]
>
> — *Iwo Jima: Amphibious Epic*
> Historical Branch, HQMC

Then, RCT 21 was ordered to swing southeast to envelop the Japanese positions holding up the 9th Marines' advance.

Hills 331 and 362-C at the Middle of Third Defense Belt

After a 200-man counterattack through a tiny gap between 9th and 21st Marines was repelled, the two regiments resumed their mutual push towards Hill 362-C on 4 March. Making little progress with the "overwhelming-firepower" approach to warfare, they rested until the 6th. On that date, an hour-and-a-half-long naval gunfire and artillery bombardment again failed to limit the amount of enemy fire. So, Gen. Erskine decided to try something completely new on Iwo—a pre-dawn attack without any preparatory fire or radio traffic. In a trick straight out of Mao's playbook, he added the following directive: "No one was to fire until it became certain that the enemy had discovered the main body of the assault force." His maneuver plan was equally ingenious. 21st Marines was to advance on 362-C, while 9th Marines moved forward only 200 yards as a diversion. Then, the good general may have received a little Providential help. As the attack kicked off at 0500 on 7 March, the night was still inky black and its light drizzle soothing. After not doing any of the preliminary reconnaissance that would have alerted enemy scouts to the early morning move, 3/9 [now attached to 21st Marines] instead relied on what later came to be known as "recon pull." And this was only after the objective had been briefly backlit by a star shell.[32]

As a direct result of all this unconventional activity, 3/9 would end up running a very successful pre-dawn attack on Hill 331 (after being incorrectly briefed it was 362-C). That assault was launched promptly at 0500 after a few smoke rounds had also been directed at the objective. For 35 minutes, 3/9's assault troops completely surprised the opposition. Then, an enemy machinegun had to be silenced by a Marine flamethrower. From that point forward to the top of Hill 331, 3/9 would encounter only scattered resistance. Most would be handled by flamethrower men who were not yet discernible as such.

Then at 0715, after a 10-minute artillery barrage of the correct objective, 3/9's CO dared to move from Hill 331 to a place even farther behind enemy lines. This time, all Japanese forces in the area were awake. Progress over the next 250 yards would be much slower than before. Many caves and bunkers would have to be neutralized, one at a time. Each would take a sub-unit maneuver. Finally at 1400, Hill 362-C was declared secure.[33] It had been a good day for 3rd Division, because of its greater reliance on the element of surprise.

> Most notable in the night attack was the fact that, although nearly all the basic dope was bad, the strategy proved very sound, since it turned out that the open ground taken under cover of darkness [on the way to Hill 331] was the most heavily fortified of all terrain captured that day, and the enemy [soldiers] occupying this vital ground were taken completely by surprise (actually sleeping in their pillboxes and caves) [9thMar SAR, Encl. C, p. 11].[34]
>
> — quote from Lt.Col. Harold C. Boehm (3/9 CO)
> *Iwo Jima: Amphibious Epic*
> Historical Branch, HQMC

But, more trouble lay ahead for soon to be famous 3/9. While it—as an attachment of 21st Marines—had moved slightly southward, the rest of 9th Marines had gone east under the cover of darkness. By daylight, the 2nd Battalion on the left and 1st Battalion on the right had penetrated the planned 200 yards into an astounding maze of enemy fortifications. Then, all the defenders of those fortifications came suddenly to life. 1/9 had to fight hard just to reestablish any freedom of movement behind "enemy lines." It would eventually break out, but 2/9 require more time to free itself.

Cushman's Pocket

Before long, Lt.Col. Cushman's 2nd Battalion, 9th Marines was taking heavy fire from front, flanks, and rear. For all practical purposes, it was cut off and trapped. American armor could not get through to help. At dusk on 7 March, elements of 2/9 were still pinned down. It would be a full 36 hours before the first two companies of that battalion would extricate themselves from the encirclement.[35]

What then came to be known as Cushman's Pocket had been on high ground. That made it a likely bastion in the enemy's Third Defense Belt. (See Map 11.6.) After 3/9 had left Hill 362-C and headed for the eastern coast, the rest of 9th Marines was still bogged down with what turned out to be the last stand of Col. Nishi's 26th Tank Regiment. Supported by well-dug-in armor with 37mm or 47mm guns and separate antitank guns, Japanese soldiers were now fighting from caves and emplacements within a cluster of sandstone buttes. No conventional U.S. attack (however well supported) could dislodge them. Making any forward progress whatsoever took the impromptu cooperation of tiny, semi-autonomous teams of American infantrymen. Each had to move from one little butte to another in its assigned lane, fight "Indian style," and then blow up any gun site or sub-surface entry point it discovered.[36]

The outlying positions of this likely fortress in the Third Defense Line may have been initially encountered from the west by 3/21 around 9 March. That was before 2/9 got trapped inside. The complex's final reduction would take an envelopment from the east end. While U.S. tanks were involved, they did not single-handedly capture Cushman's Pocket. That would take the cooperation of many infantry fireteams.

> [On 9 March,] the 9th Marines, with 3/21 attached, hit a stone wall of resistance. Even though tank support was available, the terrain severely limited the employment of armor. The enemy was not slow to take advantage of this situation. He first fired a number of air bursts over one of the tanks in order to disperse the infantry. Once this had been accomplished, he dispatched a demolition detachment under cover of a smoke screen which put the tank out of action with a demolition charge and a Molotov cocktail.

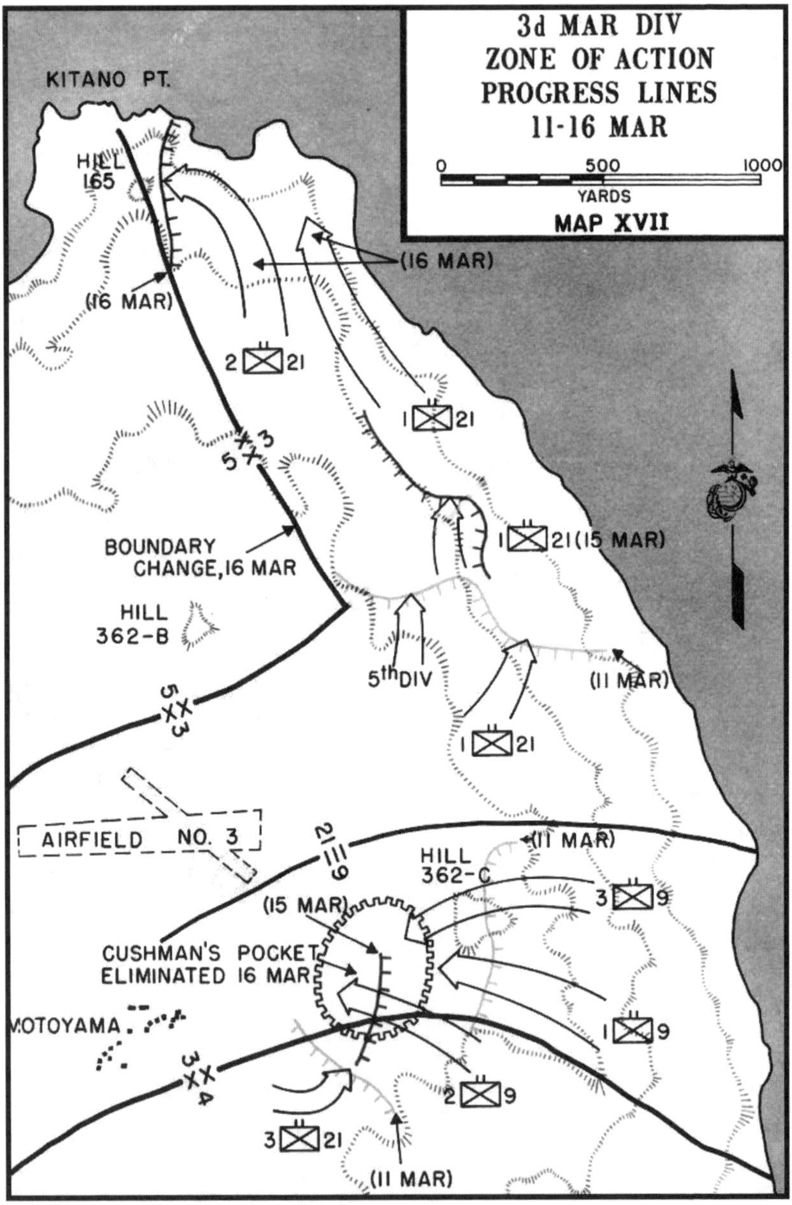

Map 11.6: Cushman's Pocket Was Also Nishi's Last Stand

(Source: Lt.Col. Whitman S. Bartley, "Iwo Jima: Amphibious Epic," Washington, D.C.: History Branch, HQMC, 1954, map XVII.)

> Neither 2/9 nor 3/21 were able to score any sizable gains during the day.[37]
>
> – *History of Marine Corps Operations in WWII*
> Historical Branch, HQMC

> The 1st Battalion, 9th Marines, which had . . . on 11 March . . . intended to sweep the last hostile troops from the ridge in the southeast corner of the 3d[Division]'s zone and make contact with 3/9, which was moving south along the coast against the same high ground. Then both battalions would turn west to attack Cushman's Pocket from the rear.[38]
>
> – *Iwo Jima: Amphibious Epic*
> Historical Branch, HQMC

When the Defense Belt bastion was finally attacked from the east by the entire 9th Marines with its flame tanks blazing, there would be no officially recognized infantry bravery. But, tiny front-line forays had still been necessary because of the labyrinth of tunnels beneath the buttes.

> Marine engineers and infantrymen continued destroying pillboxes and caves during the hours of daylight, but at night the enemy crawled out of other caves and holes to hurl grenades and then dart back into their hideouts.[39]
>
> – *Iwo Jima: Amphibious Epic*
> Historical Branch, HQMC

> By morning of 11 March the only effective opposition to the 3d Division came from terrain along the right boundary, where the Japanese held a . . . stubborn core of resistance [1000 yards] southwest of Hill 362-C. This last strongpoint, which had already withstood several days of attack by 1/9, 2/9, and 3/21, was named "Cushman's Pocket," after the commanding officer of 2/9 [who had first gotten into trouble there]. . . .
>
> On 12 March, the 1st and 3d Battalions [of the 9th Marines] attacked west toward Cushman's Pocket while 3/21 acted as a holding force on the opposite side. On the right (north) flank 3/9 had relatively easy going in its zone of action, but 1/9 met stiff opposition and moved slowly. An armored bulldozer cleared a tank road to the 1st Battalion's

front, but at no time could more than two Shermans operate together in the rough terrain. Infantrymen [in tiny contingents] worked their way forward among the miniature sandstone buttes to guide tanks and point out the camouflaged enemy emplacements. Most of these positions were so well concealed that they escaped detection until Marines were almost on top of them. As usual in this type of fighting, the infantry flame throwers and demolitions proved the most successful weapons.

The 3d Battalion, 21st [Marine Regiment] . . . on 13 March . . . left its position on the west side of the shrinking pocket. The 1st and 3d Battalions, 9th Marines continued pressing the attack from the east. On the north flank, 3/9 completed reduction of that portion of the pocket in its zone of action, and 1/9 made substantial gains to compress the enemy-held area to about 250 yards on a side. Regular Shermans and flame tanks operated very effectively on the 13th, neutralizing a number of concrete emplacements. Upon completion of its assignment against the pocket, the 3d Battalion, 9th [Marines] was ordered to the Hill 362-C area where it undertook a systematic mop-up [because all Third Defense Belt locations were probably connected by tunnel].

The next day, 14 March, operations against Cushman's Pocket entered the last phase. The 1st Battalion pressed the attack during the morning and early afternoon to gain about 100 yards. Then at 1530, the 2d Battalion passed through that unit to finish the job. A large capacity flamethrower tank, on loan from the 5th Division, and Shermans from the 3d Tank Battalion scorched and pounded the last enemy defenses until only sporadic resistance challenged infantrymen as they moved in to finish the task.[40]

— *Iwo Jima: Amphibious Epic*
Historical Branch, HQMC

Riflemen had once again guided in flame tanks and consolidated burned areas. Thus, fireteam maneuver became undeniably the key to decisively beating a well-dug-in foe.

What might have turned into a routine chore of mopping up turned instead into a rather protracted operation requiring all the skill the attacking force could muster. The action that 2/9 saw for the remainder of 14 March was subsequently described with the statement: "Inconclusive hand to hand fighting ensued until dark (2/9 AR, p. 3)." [41]

– *History of Marine Corps Operations in WWII*
Historical Branch, HQMC

The Increasing Importance of Tiny-Element Stratagems

As the Japanese began to use northern Iwo's jumbled terrain for their "last-ditch efforts," a trend (since landing) within U.S. small-unit tactics became more pronounced. Extensive close-quarters combat had been required at the craggy center of Hill 362-B. And, according to the 5th Marine Division website, "the number of caves had increased in the island's north, requiring more demolitions."[42]

The narrow confines of the Gorge had made more fireteam coordination necessary, as had the sandstone maze of Cushman's Pocket. U.S. flame tanks and indirect fire could only partially reach both locations, so full credit for their final reduction does not belong with American technology. Fancy weaponry had also been available during the first few days of the invasion. That the U.S. casualty count had gone way down during the last 22 days of the 36-day ordeal (only 11,000 of the 33,000 total) is clear evidence that the Marines' short-range fighting skills had improved.[43] This was true, despite the 3rd Marine Division commander's observation that the replacement draft needed more training in "small-unit teamwork" and "cover and concealment."[44] 3/9's breakthrough of the Third Belt had been unexpectedly successful. So, Gen. Erskine may have been simply acknowledging the extent to which fireteam skill contributed.

This chapter's accounts of Third-Belt battles reveal more grenade work than before. But, that for Cushman's Pocket also talks about "Indian-style" combat (that which better preserves the participant's life [45]). Within it, "blasting away" at the foe probably refers to the systematic use of handheld explosives.[46] Many grenades had been used atop 362-B.[47] And, in both places, the enemy had been too near for bazooka work.

So, at the heart of fireteam maneuver had been getting close enough to an enemy position to toss in a grenade. It also took something a peacetime military regularly ignores—how sister buddy teams can alternately cover and move in various kinds of microterrain. In today's world of high-tech gadgetry, such mundane field exercises regularly take a back seat to new-gear classes. But, the life or death struggle for Iwo had demanded their impromptu development, despite the high turnover of personnel. Such development requires the initiative of the least experienced members of an infantry outfit—despite organizational pressures to overcontrol them for safety's sake.

12 During the Final Consolidation

- Where were enemy troops found after the Marines had left?
- What finally happened to Col. Nishi?

Many Japanese apertures had gone undiscovered.

(Source: Charles Waterhouse painting, as part of "Closing In," by Col. Alexander, Marines in WWII Commemorative Series, HQMC pamphlet, 1994.)

Resolving the "Final Resistance" at the Gorge

The most likely spot for a fake last stand on Iwo would have been a narrow canyon far from the actual fallback position. Then, to make sure adequate attention was paid to the decoy site, one would erect a domed communications center nearby and fight harder than expected for the whole area with a small, but highly skilled, rear guard.

If unscathed enemy soldiers had been falling back from one defense line to the next, they would have done so through the island-long tunnel in the Japanese documentary. (See Map 12.1

and then look back at Figure 2.9.) All wounded would have taken the same route. Not far from its crossing of the Third Defense Belt, the U.S. Army would find an underground hospital of the 2nd Mixed Brigade. But, that Brigade had been based at Turkey Knob and mostly defending the Second Belt.[1] That probably means all injured Japanese soldiers had been shifting toward the island's northeastern end.

Kuribayashi's intermediate headquarters at Motoyama Village lay 1000 yards north of Turkey Knob, and squarely atop the island-long tunnel.[2] Not until 4 March did the Japanese commander move his headquarters from "the central highlands" to the Gorge cave.[3] His most likely route would have been up the island-long tunnel and then left along the Third Belt's own underground passageway. It was this multi-level Gorge CP that the U.S. Army would later find to be mysteriously reoccupied.

> Another incident, on 4 June, involved the location and identification of the headquarters cave reputedly used by the island's commander, Lt.Gen. Tadamichi Kuribayashi. Although the 147th Infantry had already identified and exploited several other large underground complexes, this one, located on the island's northeast quadrant, was the largest of all.[4]
>
> — "The 147th Infantry Regiment at Iwo Jima,"
> U.S. Army Center of Military History

The general's remains would never be found anywhere on the island. Neither would Iwo's expansive tunnel network ever be fully investigated by the Americans.

The "Last-Ditch" 26 March Attack

Japanese troops were seen exfiltrating the cordon around Kuribayashi's Gorge cave after dark on 25 March. But, they may not have been the same 300 who later appeared inside the Airfield No. 1 billeting area. Must the POW claiming Kuribayashi had led this attack be believed?

Near the end, rear-guard Asian troops can be very deceptive. Others with direct knowledge of the sub-surface passageways at Airfield No. 1 likely conducted the late March attack. Another

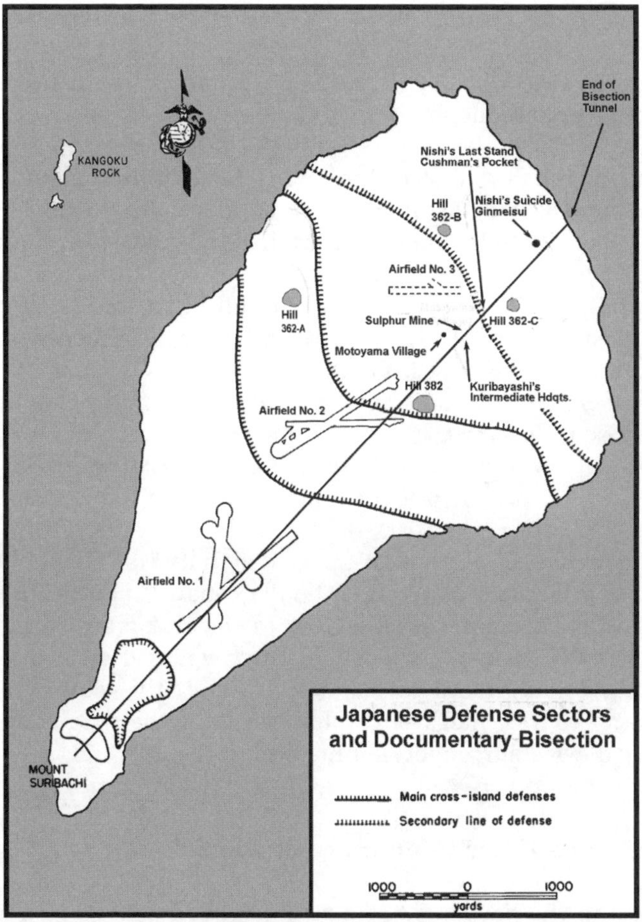

Map 12.1: More Events Where Documentary Had Bisected Island
(Source: Alexander, "Closing In: Marines in the Seizure of Iwo Jima," Hist. & Museums Div., HQMC, 1994, pp. 5, 8.)

event would soon demonstrate how much damage a few bypassed troops can do through an intact tunnel system. For some reason, the 147th Infantry Regiment's consolidation chronicle fails to mention this rather dramatic occurrence.[5]

The Japanese Tunnel Network Had Been Undisturbed

Enemy resistance on Iwo Jima did not end with the airport raid and Gorge reduction.

> During the first week of April, in an incident reminiscent of the unexpected enemy attack of 26 March [on Airfield No. 1], about 200 Japanese materialized just above the East Boat Basin, where they attempted to rush an infantry command post. This battle continued all night and all of the attackers were killed, but not before they had succeeded in exploding 6,000 cases of [U.S.] dynamite (HQ, Army Garrison Force G-2 Weekly Rpt No. 2, dtd 8 Apr 45, as cited in Morison, *Victory in the Pacific,* pp. 69-70.)[6]
>
> — *History of Marine Corps Operations in WWII*
> Historical Branch, HQMC

The reason for all this unexpected activity was simple. Kuribayashi had planned all along to finally conduct a guerrilla war in which all subordinates would fight to the last man. And, there were many "last men" still left in tiny bypassed enclaves—most linked by underground passageway. Many still wanted to fight. Others, of course, had been wounded. But, the code of *Bushido* only disallowed surrender, not medical evacuation.

Where Were Those Remaining Enclaves?

Late in the battle, U.S. forces had relied heavily on flame at several key locations: (1) Turkey Knob;[7] (2) Senda's last stand;[8] (3) Cushman's Pocket;[9] and the Gorge cave.[10] But all four spots possessed enough of an underground network to allow for survivors. The people who attacked just above the East Boat Basin in April could have come from any of the large Meatgrinder groups (some 1000 yards away through a combination of First and Second Defense Belt tunnels), or a mixture of holdouts from other places. Few, if any, tunnels had been searched during the 36-day fight. That's because they had likely been boobytrapped to preclude discovery of lower levels. The Marines had been having such a tough time at the surface that they were in no hurry to risk more casualties below ground. So, still unaware of the extent of the sub-surface labyrinth, they tried to negate its effects through a myriad of surface portal closures.

> Many caves, tunnels, and other underground positions were sealed up by our forces without being entered.[11]
> — *Intelligence Report,* G-2, TF 56, April 1945

147th Infantry Only Found a Big Bunch in Two Places

The first holdout location to be discovered was the 2nd Mixed Brigade Hospital. It was on the next belt back from the one that this Brigade had been manning.

> Another significant event occurred on 11 April when an eleven-man patrol from Company A . . . took the largest number of prisoners at one time during the entire battle of Iwo Jima near Target Area 202-Fox, located on the eastern portion of the island. When the patrol spotted two Japanese soldiers emerging from a hole near its ambush position during the early morning hours, Kolb's men opened fire, killing one and seriously wounding the other. Despite his wounds, the Japanese soldier managed to crawl back into the hole, prompting Kolb to use his Nisei interpreter, Sgt. Ritsuevo Tanaka, to call to the Japanese thought to be still underground to come out and surrender or they would be sealed up alive with explosives. After a brief negotiation, Kolb learned that his patrol had stumbled upon the hospital of the 2d Mixed Brigade, located 100 feet underground.[12]
> — "The 147th Infantry Regiment at Iwo Jima,"
> U.S. Army Center of Military History

The FMFPac G-2 map in the bibliography shows what existed at this alphanumeric location. It was about 1000 yards east of Motoyama Village and 1000 north of Higashi. That puts it very near the Naval Headquarters bastion on the Third Belt that Naval Forces had tried to reach in their big counterattack. It was also only 300 yards from the island-long conduit.

The second big spotting was apparently at Kuribayashi's Gorge headquarters. But, its precise location hadn't been reported. Neither did it have as happy an ending.

> Another incident, on 4 June [1945] , involved . . . the headquarters cave reputedly used by . . . Kuribayashi. . . . Discovered by a patrol from Company F, led by Lt. James I. Ahern, the cave was found to be still defended by the enemy, who refused calls to surrender. Calling forward a demolition team from the regiment's Ammunition and Pioneer Platoon . . . , the soldiers pumped in hundreds of gallons of gasoline and set it alight. The resulting fire ignited a quantity of ammunition stored inside, killing or wounding many of the surviving occupants who had not committed suicide. Fifty-four survivors surrendered, though two killed themselves shortly afterward.[13]
>
> — "The 147th Infantry Regiment at Iwo Jima,"
> U.S. Army Center of Military History

Holdouts Were to Emerge from Iwo's Caves for Four Years

On 6 January 1949, Yamakage Kufuku and Matsudo Linsoki, two Imperial Japanese Navy machinegunners, surrendered on Iwo Jima.[14]

What Had Become of the Mysterious *Ginmeisui?*

By some accounts, Col. Nishi had received a blinding wound at Cushman's Pocket. He then committed suicide at either "Ginmyosui" (the *Ginmeisui* of Chapter 1) or "Futagoiwa."[15] (See Map 12.2.) There's a human-head-shaped rock named Futagoiwa along Nagasaki's picturesque waterfront. So, like *Ginmeisui*, this other location may have simply had something to do with memories of home.

Or, that second destination had actually been *Futago iwa*—futago meaning "twin," and iwa meaning "rock." There had been two nearly identical offshore islets where the documentary bisection reached Iwo's northeast coast—(GC) 806760.[16]

Newcomb's book mentions the Iwo location of *Futatsuno iwa*—meaning twin rocks in Japanese and called "Fatatsune Rocks" by the Americans.[17] In 1992, the U.S. Coast Guard Loran Station would only acknowledge the existence of a single Fatatsune Rock (whereabouts unknown). But, *Futatsu iwa* most directly translates as "notification rock."[18] Only under an old Japanese counting system, does its apparent opposite—*Futatsu no iwa*—mean "pair of rocks."[19] A close look at Iwo Jima's immense shoreline on the detailed G-2 map reveals a nearly identical pair of good-sized rocks at only one place—GC 806760.[20] So, *Futatsuno iwa* may have been different from Futatsu Rock (two outcroppings of unequal size between Red 1 and Red 2 beaches). And the Nishi story would then lend credence to *Ginmeisui* and *Futago iwa* (a.k.a. *Futatsuno iwa)* both being waystations on a highly secretive evacuation route.

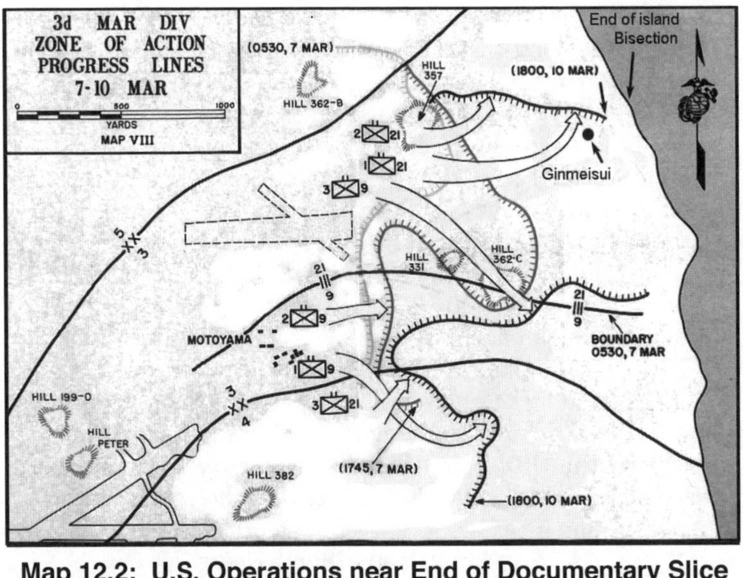

Map 12.2: U.S. Operations near End of Documentary Slice
(Source: Lt.Col. Whitman S. Bartley, "Iwo Jima: Amphibious Epic," Washington, D.C.: History Branch, HQMC, 1954, map VIII.)

More About the End of that Island-Long Passageway

Around 10 March, several openings were discovered in the cliff line along Iwo's northeastern coast. (See Map 12.3.) The U.S. Navy fired at them and 21st Marines temporarily occupied the bluffs above them in its initial drive on Kitano Point. There is even a photograph of such an opening in the most prestigious battle chronicle. Its caption reads "landslide blocked . . . entrance to a cave." But, this so-called cave had a perfectly round mouth over 15 feet in diameter that had been only partially blocked by shelling debris.[21] At this stage of the battle, 3rd Division units had been searching for gaps in the Japanese defense system (following paths of least resistance). After very little trouble reaching the above mentioned coastline, its units would not have taken the time to search cliff-face openings.

> Continuing the drive to the sea, both 3/9 and 1/21 sent patrols down over the cliffs to the beach. [Look back at Map 11.6.] Resistance to the front was sporadic and easily overcome. An airstrike was called in against one stubborn enemy group, and . . . a direct support destroyer directed its fire into caves and emplacements along the cliffs.
>
> Although Company A (1/21) was the first tactical unit to reach the northeastern beaches, a reconnaissance patrol led by the battalion intelligence officer arrived earlier. . . .
>
> The 2d Battalion, 21st Marines, was still unable to push its left (north) flank [of the formation] ahead to the edge of the cliff because of the enemy resistance, . . . but on the right it moved forward, keeping contact with 1/21. Although patrol activity all the way to the waters' edge had revealed that enemy resistance was light, all battalions held their positions on the bluffs above the beach for the night. . . . [Look back at Map 12.2.]
>
> . . . [I]n the 21st Marines' sector, patrols from 1/21 established the fact that there were no enemy on the beach, and that unit swung to the north with its right flank moving along the cliff[top]s. By 1500 [the next day] all organized resistance had been eliminated in the 1st Battalion's zone of action.[22]
>
> — *Iwo Jima: Amphibious Epic*
> Historical Branch, HQMC

To keep from being discovered, the thousands of wounded Japanese troops who may have occupied the tunnel-end chambers in Figure 2.9 would have only had to limit surface forays. Due to a much longer battle than expected, there was undoubtedly pressure from TF 56 for all ground units to conclude their above-ground sweeps.

> With its lines tied to Company I, 27th Marines, . . . [21st Marines] held positions along the cliffs overlooking the coast and sent combat patrols out along the beaches below. The advance of the 5th Division governed progress along the bluffs and shorelines.
>
> Marine engineers and infantrymen continued destroying pillboxes and caves during the hours of daylight, but at night the enemy crawled out of other caves and holes to hurl grenades and then darted back into their hideouts. Alert Marines shot and killed many of these night raiders before they could make good a retreat.
>
> By 15 March, the 1st Battalion, 21st Marines had extended about 600 yards northwest-ward into the 5th Division zone and held positions along the bluff and outposts down on the beach.[23]
>
> — *Iwo Jima: Amphibious Epic*
> Historical Branch, HQMC

Initial U.S. intelligence reports concerning this area of coastline had concluded it would not permit a big landing. But, they did notice an anomaly of sorts at one specific location. This is also the place where the Japanese had strung their only barbed wire.[24] (See Map 12.4.)

> Lack of noticeable activity in the north led intelligence officers to assume that enemy fortification in that locality held less attention, and for good reason. There, nature allied herself with the Japanese. The steep cliffs would all but prohibit invasion over northern beaches. Even so, one [Japanese] battalion was reported to be in defensive position along this northern coastal area with its companies deployed along the rocky heights overlooking the ocean.[25]
>
> — *Iwo Jima: Amphibious Epic*
> Historical Branch, HQMC

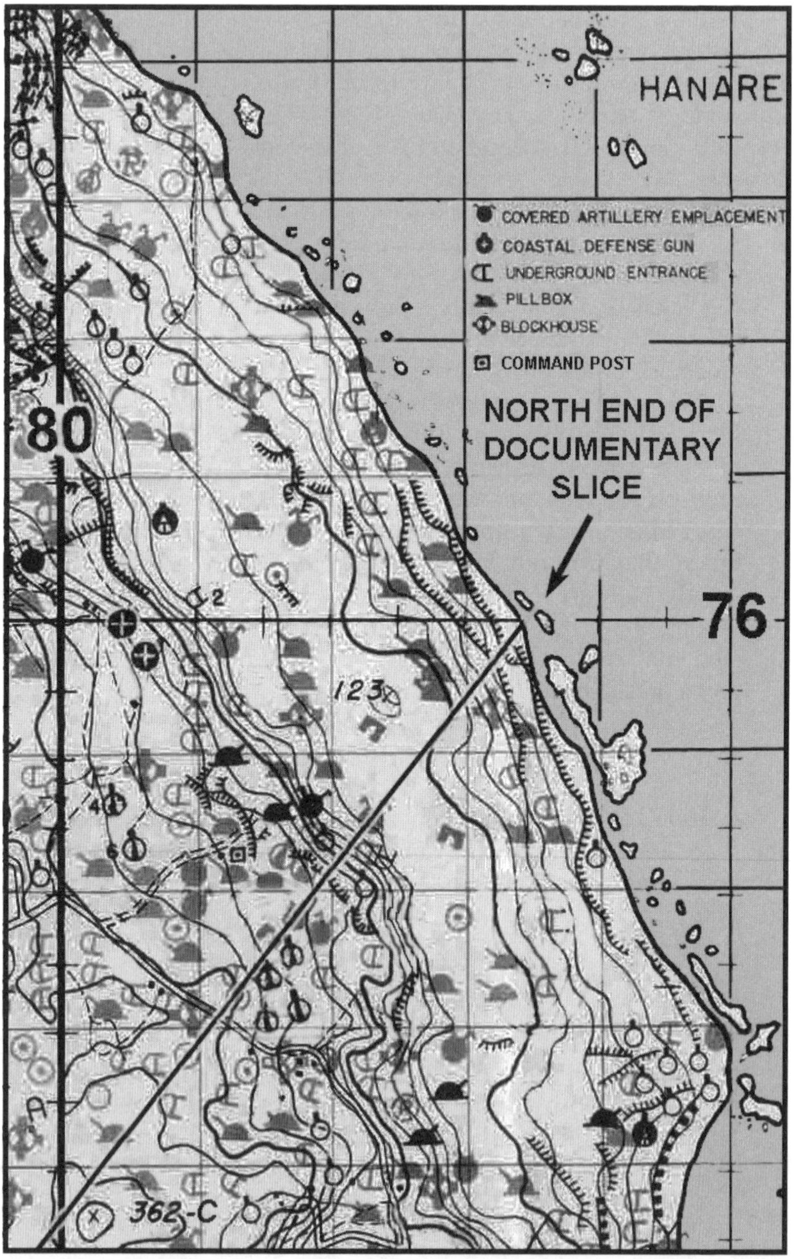

Map 12.3: Where NE Iwo Was Bisected by Cameraman Special

(Source: G-2 FMFPac "Report of Intelligence," map from this url: http://www.ibiblio.org/hyperwar/PTO/Iwo/ComTaskForFiftySixIwoJima/TaskForFiftySix-EnclosureC.pdf.)

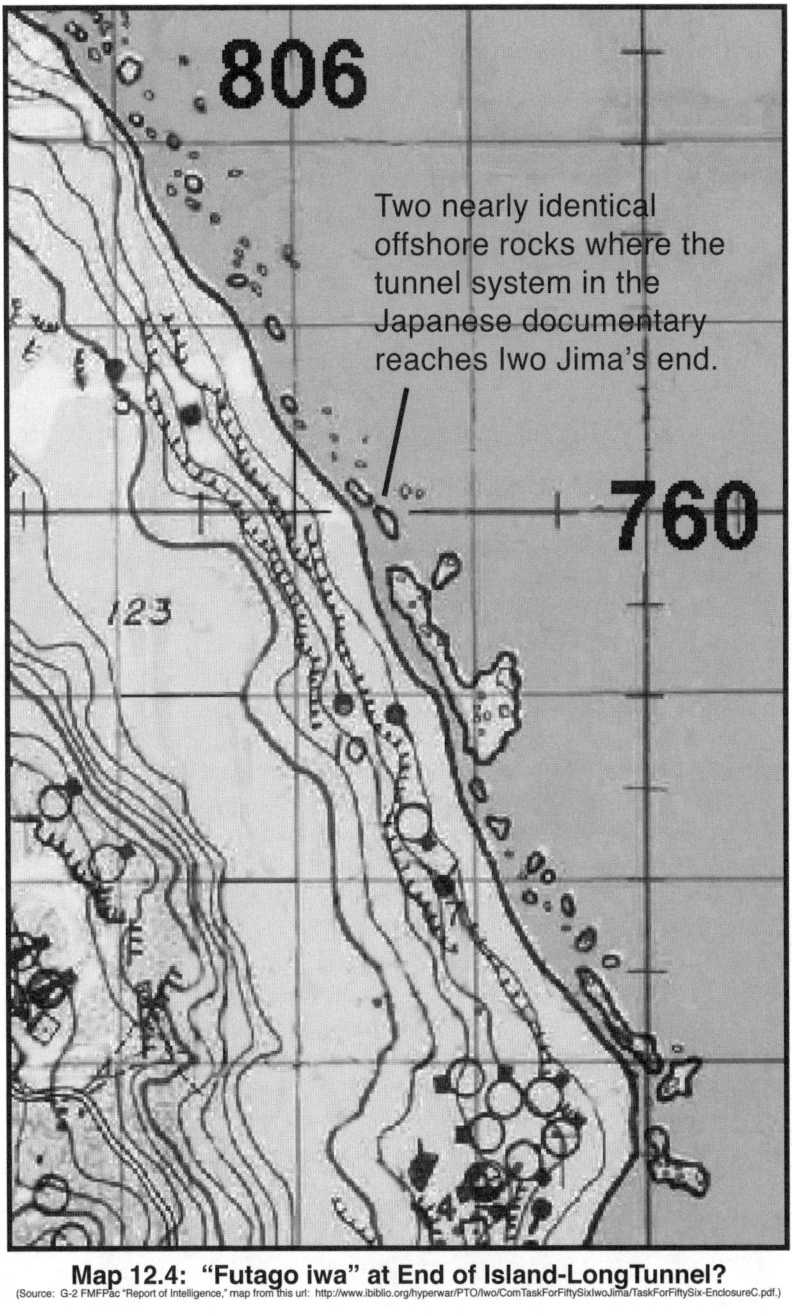

Map 12.4: "Futago iwa" at End of Island-LongTunnel?

(Source: G-2 FMFPac "Report of Intelligence," map from this url: http://www.ibiblio.org/hyperwar/PTO/Iwo/ComTaskForFiftySixIwoJima/TaskForFiftySix-EnclosureC.pdf.)

Particular attention should be paid to Hill 123. *Ginmeisui* may have been nearby. What look like barracks on either side of the island slice are actually just "open artillery emplacements," according to the G-2 map legend. But, there is an interesting CP some 2000 yards southwest of Hill 123. The darker map symbols denote enemy emplacements spotted from the air prior to D-Day, while the lighter ones are from aerial photographs after D-Day. So, two coastal defense guns had long guarded where the documentary tunnel ended.

An earlier G-2 map shows another Japanese CP on the machinegun-infested promontory at the right bottom of Map 12.4—around GC 809753.[26] Because there had been so little chance of a U.S. landing at any place along Iwo's rugged northeast coast, one has to wonder why the enemy had so carefully protected this tiny part of it.

PART FIVE

The Gift of "Improvise, Adapt, and Overcome"

"THEN I HEARD THE VOICE OF THE LORD, SAYING, 'WHOM SHALL I SEND, AND WHO WILL GO FOR US?" THEN I SAID, "HERE AM I; SEND ME."
— ISAIAH, 6:8

(Source: Attributed to "New American Standard Bible.")

13 BEST WAY TO BREACH A MODERN DEFENSE "LINE"

- What constitutes a modern defense line?
- Why can't a fully supported armored brigade penetrate it?

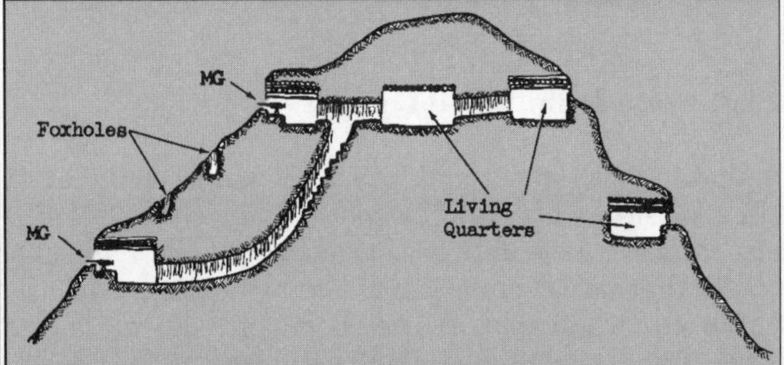

Fig. 8. Cross section of hill, showing machine-gun emplacements

A 1950's ridgeline strongpoint with reverse-slope option.

(Source: Excerpt from Chinese Manual on Field Fortifications, in "A Historical Perspective on Light Infantry," U.S. Army Combat Studies Inst., Research Survey No. 6, p. 88.)

Iwo Had Been a Major Learning Experience for America

As the Marines came to better understand their dire circumstances on Iwo, they changed how they operated. One can tell from the casualty count. After one day ashore, they had 2400 killed and wounded. At the end of the 36th day, they had 32,000. If their combat methods hadn't been altered, that end total could have been 86,400 (36 times 2400). And most of the tactical refinements would come while assaulting Suribachi and the first two defense lines. Upon moving on to the third, the island's new occupiers would suffer only 11,000 more casualties over the last 22 days of fighting.[1]

This chapter will look for what U.S. outfits did differently as the battle progressed. Some of it had to do with maneuver, some with equipment, and some with on-the-spot improvisation by frontline personnel.

By D+14, most of the seasoned veterans of all three divisions had been injured. So, members of the "under-trained" replacement draft may have had more to do with the final victory than history records. There were also untold numbers of non-infantry Marines suddenly transferred to the grunts. While senior officers worried about lack of experience in the ranks after D+14, the casualty rate actually started to fall. Flame tanks would get the majority of the credit, but much of it may belong to recently joined "hard-chargers."

Iwo's Yet-to-Be Assimilated Lesson

On that tiny volcanic island, U.S. forces discovered that—when facing a state-of-the-art rural defense—a tiny element with a few grenades can make more initial headway than a large, fully supported unit. With that initial probe can also come a lower final casualty count. The reason is simple. In littered though relatively flat terrain, one or two grunts can often outflank an enemy machinegun, while their buddies create a diversion. All it takes is a little crawling. (That's why many of Iwo's heroics so closely mirror the short-range infiltration techniques in *Strategic Rifleman.*[2]) But, this paradox of modern war has been rarely mentioned by U.S. historians. Most have focused on recently developed equipment.

> [A] small unit, a squad or even a fireteam, that is properly trained in modern, post-machinegun techniques can be just as effective as a much larger unit, while offering the enemy fewer targets. The German Army, which excelled in drawing lessons from its combat experiences, found as early as World War I that the only difference between a squad attacking a machinegun position and a company doing so was in the number of casualties suffered. Not surprisingly, by 1918 the *Stosstrupp,* a squad-sized unit, was the basic German tactical building block. In contrast, in most Marine infantry units today, the squad is regarded as merely a subset of the platoon, seldom trained for independent action. The result, in combat, is likely to be a lot of dead Marines, Marines whose

deaths could have been avoided if tasks were assigned to smaller units.[3]

— W.S.Lind, Maneuver Warfare Handbook author (in *One More Bridge to Cross* Foreword of 1998)

Because tiny contingents can more easily sneak up on a skilled adversary, elite Western commandos (like from Britain's Special Air Service [SAS]) now operate in teams of four. But, Asian armies have for millennia relied on lone infiltrators.[4]

Effect of the Mid-20th Century on Modern-Day Warfighting

The 2020 Posterity Press release, *Super-Squad,* contains detailed evidence that the current state of the art for small-unit infantry tactics only in the Eastern World until about 1975. Then, it began to atrophy in China and Russia through too much emphasis on technology. In fact, that study shows all big Eastern armies (the IJA included) using almost identical tactical techniques near the end of WWII for common combat scenarios.[5] That makes what the Marines had to learn the hard way on Iwo Jima doubly important to winning future conflicts.

Despite the ongoing advances in technology, a modern-defense-line penetration attempt would still encounter three problems: (1) diminished headquarters control; (2) serious attrition of armor; and (3) the failure of artillery and airpower to meet expectations. Near any enemy MLR, U.S. units (however well armed or motorized) would be unable to advance. Any forward motion would be by elements too small to be noticed—like a few crawling infantrymen. In other words, as in 1945 on Iwo Jima, that penetration attempt would have to be preceded by tiny, 3GW-like probes. The exact nature of those probes will take another quick look at history.

Kuribayashi's Interruption of U.S. Headquarters Control

Within the well-researched film "Battle Cry," a commanding U.S. general admits to much of the planning for a WWII invasion "going out the window with the first shot fired."[6] That's because the opponent was an active participant—something that 2GW tends to forget. The Iwo landing provided Marine units with their first full-blown example of this unfortunate reality. There, they faced highly opportunistic 3GW squad stratagems that have—to this day—remained "state-of-the-art" for similar circumstances. All those defensive pullbacks and spoiling attacks had been the brainchild of a "bottom-up-oriented" Asian commander.

From one hour after hitting the beaches until far into the 36-day struggle, American outfits all over the island would be subjected to pre-registered indirect fire. Near any of the contested ridgelines, they also encountered a veritable wall of small-arms bullets. In other words, all major maneuver elements became "pinned down," and their tactical options limited.

Soon, all divisional, regimental, and even battalion commanders had little choice but to delegate most tactical-decision making to rifle company leaders. They, in turn, were forced to rely upon individual volunteers. One or two normally enlisted personnel would conduct tiny forays to search out and destroy offending machineguns. Until they did, their parent unit couldn't advance without heavy casualties.

As the battle progressed, most U.S. headquarters elements no longer focused on operations or tactics, but only on combat support—e.g., naval gunfire, artillery, close air, resupply, reinforcement, or evacuation. Staff officers still provided the lanes up which battalions would move. On occasion, one battalion might even be allowed to enter another division's zone of responsibility. But, the overriding concern for those "in charge" had become the provisioning of frontline fighters. What little rolling stock they had left attracted artillery fire, so they often resorted to working parties.

While high-level U.S. commanders would have preferred all subordinate units to be on line and well-tied in after dark, not all required it. Their absolute control over all sub-unit activity had (through battlefield necessity) suddenly ceased to be helpful. And with its absence would come enough leeway and flexibility for frontline personnel to keep their units moving. This was a major departure from what normally occurs in a stratified Western bureaucracy (everyone doing exactly as told). It was to give new significance to the old Marine adage—"improvise, adapt, and overcome." That adage may have originally had to do with compensating for hand-me-down equipment. But, its true worth would emerge in frontline combat. Without the initiative of individual riflemen on Iwo, the divisions would have taken far more casualties. Just as most

Asians understand from their bottom-up culture, the Americans finally realized that collective effort could outperform headquarters control.

The Ensuing (But Only Temporary) Paradigm Shift

The three-divisional force that had hit Iwo's southern beaches in 1945 had almost certainly intended to move around the island in battalion size with all weapons blazing. But, each rifle squad had also been prepared to tackle a bunker, so that squad was to be the smallest cog in the wheel. Unlike the big Eastern armies of this era, the USMC had yet to acknowledge that a single fighter could make a strategic difference.

Then, this massive American juggernaut was met with such fierce resistance that full squads were being decimated by a single shell. And in the initial fighting above the Rock Quarry, whole companies were "being cut to ribbons" by automatic-weapons fire.[7] To limit casualties, many U.S. commanders were soon preceding all contested assaults with machinegun-hunting infiltrators. Unbeknownst to those American COs, so doing amounted a more advanced style of warfare. But, their "by-the-numbers" bunker-busting techniques of 2GW had become no longer viable due to the streams of bullets now crisscrossing the front of every objective. The U.S. company's only other way of advancing was a ragged line of semi-autonomous and fully cooperating fireteams. While some buddy teams crept up on the forwardmost foe positions, others suppressed sources of diagonal fire. Such a maneuver would work well for a former Raider battalion on Okinawa's Shuri Line in May of 1945,[8] but it has no known counterpart on Iwo Jima. As a result, most TF 56 rifle company COs understood only the preliminary probe option.

The Obvious Had Been Under-Appreciated

As with the special-operations forays of today, this new style of warfare required deeply decentralized control. The Iwo rifle companies had been so thoroughly pinned down that they were unable to cover each other's movements by fire. Nor could their platoons or squads. That's when low-ranking individuals would have to be trusted with each unit's future. Until now, recent enlistees hadn't been given credit for that much skill or determination. But, desperate times call for desperate measures, and some of the "new guys" had talked of putting wild game on the table during the Depression. So, the implicit key to all battlefield success was finally being tried by the Americans—the lone gladiator. His inherent cunning could be more powerful than any weapon. Many Japanese bunker occupants had already suffered the consequences of that simple truth.

The Physical Makeup of Such an MLR

A state-of-the-art defensive belt is nearly impossible to spot until it unleashes its fury. Then, to the encroaching unit, it looks like back-to-back rows of bunkers, each slightly offset from the last (like roofing shingles). To their immediate front is an impenetrable wall of interlocking (and often grazing) machinegun fire. Any way to avoid it (like a shallow depression in the earth) has been covered by mine or preregistered mortar.

To complicate matters, such a death-dealing array can also engage targets far to its front. It is, in effect, an ambush with an extensive kill zone. So after considerable danger in the approach march, the attacking unit must close with the forwardmost machinegun nests. Those nests, in turn, protect a pillbox or subsurface chamber that doubles as a bomb shelter. But, it is the labyrinth of tunnels beneath each defensive array that makes its seizure such a formidable task. After the enemy's initial positions have been overrun, he easily conducts surprise attacks behind the intruder's lines.

Only Way Out of Such an Ambush

A Marine outfit could not easily approach the forward-most positions of such a belt on Iwo. Fully exposed tanks were in as much danger as people. Some of the lead emplacements had to be engaged from the side or back by a tiny maneuver element. It normally consisted of one or two teenagers wriggling forward through runoff ditches and rubble. Luckily, some of their exploits have been recorded in award citations. Grenades were the weapon of choice, because they were in great supply, capable of being thrown farther and more accurately by some men, and often causing a movement-concealing cloud of dust.

> 15 March 1945. Repeatedly exposing himself to hostile fire, Sergeant Marini [of F/2/25] bravely led his platoon toward its objective until enemy machine-gun and mortar fire rendered further progress impossible. Crawling alone seventy-five yards ahead of his platoon, he reduced the Japanese machine gun position with rifle fire and grenades and, subsequently discovering an enemy mortar position which had been shelling his company, advanced on it alone until he was close enough to destroy it by grenades.[9]
>
> — Navy Cross citation for Sgt. Daniel J. Marini

But, the satchel charge remained a vital part of seizing a heavy fortification. And the newly acquired bazooka helped to get close enough to insert one. Its bigger impact created blinding dust and disorienting concussion. Smoke grenades were seldom used for screening movement, because they also obscured the source of the bullet barricade. After realizing the effect of the bazooka, PFC Jacobson became only the first of many "one-man wrecking crews."

> 9 March 1945. . . . [Plt.Sgt. Julian of C/1/27] fearlessly moved forward to execute a 1-man assault on the nearest pillbox. Advancing alone, he hurled deadly demolition and white phosphorus grenades into the emplacement, killing 2 of the enemy and driving the remaining 5 out into the adjoining trench system. . . . Intent on wiping out all resistance, he obtained more explosives and, accompanied by another Marine, again charged the hostile fortifications and knocked out 2 more cave positions. Immediately thereafter, he launched a bazooka attack unassisted, firing 4 rounds into the 1 remaining pillbox and completely destroying it before he fell, mortally wounded.[10]
>
> — MoH Winner P/Sgt. Joseph R. Julian

Modern Lines of Resistance Are Supposed to Be Soft

As the Marines moved northeast from Airfield No. 1, they encountered a plateau with several hilltops. Along the ridgelines between those hilltops lay connected matrices of squad-sized strongpoints. Because the surface material was easy-to-dig volcanic "tuff," all strongpoint clusters had been linked by underground passageways. Through such passageways, hard-pressed members of each cluster could be reinforced or resupplied from within the same belt. Or, survivors of that cluster could move back—by an indirect route—to a succeeding belt.

Sometimes, enemy troops would just shift to the reverse slope of their current ridge to contest—from behind—the Marines' assault on the next ridge. That's why advancing U.S. units needed rear echelons. But, much of the upfront activity was still being conducted by GI buddy teams, as with the Marine Raiders' assault at Baikoro. (More on that later.) Large-unit sweeps still occasionally happened, but not as often as a simpler formation. In one photo, a whole unit can be seen advancing toward Nishi Ridge in a staggered column beside two tanks.[11] And after all the problems the Marines had crossing the isthmus and Rock Quarry rim, that column was likely preceded by a pair of infantry scouts. When it was finally contested, its first reaction would have been a tiny foray to locate the offending machineguns.

Reconnoitering Far to the Front of Friendly Lines

Very few of these U.S. hunting expeditions were reconnaissance patrols in the normal sense of the term. Should any (however small) happen upon a hole in the ground from the side or back, its members tossed in a fragmentation grenade.

> [O]n 7 March 1945. Painfully wounded by a sudden burst of hostile shell fire emanating from a well-concealed Japanese emplacement while reconnoitering an enemy infested area several hundred yards to the front of our lines, Private First Class Gabriel [member of 3/28 "advance patrol"] voluntarily risked his life in a valiant effort to wipe out the resistance. Realizing that he alone knew the location of the strong point, he dauntlessly crawled forward and, upon raising his head to reconnoiter again, was struck a second time by the enemy. Determined to complete his perilous mission despite his waning strength, he steadfastly main-

> tained his position, relentlessly firing his remaining rounds of ammunition and indicating the location of the emplacement before he succumbed to his wounds.[12]
>
> – Navy Cross citation for PFC Willard F. Gabriel

Civilian Aptitudes Had Suddenly Become Important

After realizing the need for individual effort, the rifle company leaders began to change how they viewed their lowest-ranking subordinates. If a single man was to save the rest, then he could be selected for certain abilities. One of the foray "volunteers" had been a center fielder at Baylor and then played minor league ball.

> 9 March 1945. . . . 1stLt. Lummus [of 2/27] slowly advanced his platoon against an enemy deeply entrenched in a network of mutually supporting positions. Suddenly halted by a terrific concentration of hostile fire, he . . . moved forward of his front lines in an effort to neutralize the Japanese position. Although knocked to the ground . . . he immediately recovered himself and . . . quickly located, attacked, and destroyed the occupied emplacement. . . . [H]e fell under the impact of a second enemy grenade but . . . staunchly continued his heroic 1-man assault and charged the second pillbox, annihilating all the occupants. . . . Held up again by a devastating barrage, he again moved into the open, rushed a third heavily fortified installation and killed the defending troops.[13]
>
> – MoH citation for a former New York Giant, 1stLt. Jack Lummus

Because 19-year-olds are generally more agile than seasoned veterans, the most junior grunts on Iwo soon wanted in on the lone-wolf adventures. Fathers had told them stories of WWI, and they wanted a few of their own. Many of these "newbies" would subsequently demonstrate their ability to lower parent-unit casualties. While a member of 2/9 near the second defense belt, Pvt. Wilson D. Watson was to earn the nickname "one man regiment." That's because he had made as much difference as three fully deployed battalions. And PFC Jacobson of 3/23 became so proficient at "one-man destruction" on Hill 382 that his services were requested by another assault company.[14] Most unit headway on Iwo was soon due to a vast series of individual exploits.

Then It Was All About Explosive Placement

The most vital step in bunker neutralization had been getting an explosive charge through the aperture. Grenades were readily available, so they were first attempted. But, for the heavier fortifications, a larger explosive was necessary. That's when special "demolition teams" came into play.

> On 13 March, when the advance of his company was held up by rifle and grenade fire from a series of caves to the front, Corporal Crabtree [of an undetermined rifle company], utterly disregarding his own safety, ran forward of the front lines and destroyed two cave positions with demolitions. Again on 15 March, disregarding enemy machine gun fire from caves to the front, he ran through the front lines and attacked cave positions with satchel charges, destroying three cave positions and making possible the advance of his company.[15]
>
> — Silver Star citation for Cpl. Luther C. Crabtree

The New Focus for Most Field Headquarters

Many infantry COs on Iwo had started to shift away from their traditional role as symphony conductor, and into one of "musician facilitator." One of the most successful battalion commanders would bypass all subunit leaders to personally coordinate key individuals.

> Lieutenant Colonel Boehm [of 3/9] . . . valiantly took over frontline leadership upon initiating a series of perilous advances on 7 March, effectively directing the coordinated efforts of flame throwers, bazookas, tanks and artillery in repeated thrusts against heavily fortified and fanatically defended positions. During four days of heavy fighting, he personally advanced his Battalion to the cliffs overlooking

> the beach at the northern end of the island, cleared the enemy from these commanding positions and then turned to operate against the Japanese forces remaining on the right flank of the original advance.[16]
>
> —Navy Cross citation for Lt.Col. Harold C. Boehm

Zippo-Tanks Had Become Most Effective Supporting Arm

While America's smaller-caliber direct- and indirect-fire weapons were having little effect on Kuribayashi's concrete emplacements, some of its Sherman tanks were making a difference. They were, of course, the ones that could spew flame up to 150 yards for 55-80 seconds.[17] But, infantry volunteers still had to show them the safest way forward. As a medium-tank variant, the M4A3 Ronson had relatively thin armor. It also carried a heavy load of combustible fluid. That made it highly susceptible to enemy damage. (See Figure 13.1.)

> 9 March 1945. Although wounded a few moments before, Sergeant Garrett [of D/2/27] voluntarily advanced ahead of the lines in the face of heavy enemy fire to guide supporting tanks in locating furiously resisting and well-concealed Japanese emplacements. Unable to make his signals understood, he further exposed himself to hostile fire by climbing on the front of the tank and entering it through the escape hatch.[18]
>
> — Navy cross citation for Sgt. Nolen M. Garrett

> 22 March 1945. Valiantly leading his platoon and supporting tanks against an enemy last-stand position which had resisted attack for seven days, Sergeant Lauderdale [of the same company D/2/27] pushed the front lines three hundred yards into Japanese territory. Advancing through a deep canyon covered by enemy fire, he requested tank support and, in the face of accurate hostile fire, moved ahead of the front lines to reconnoiter routes for the tanks and to direct their fire against the Japanese. Exposing himself on six different occasions to point out targets for the flame thrower tanks, he contributed directly to the success of the

> mission and was largely responsible for keeping the casualties in his platoon at a minimum.[19]
>
> —Navy cross citation for Sgt.Claude E.Lauderdale

While attacking up the backside of the infamous Gorge, American flame tanks came into full usage. But, they were carrying a napalm-like substance and had to be kept in hull defilade amongst all the runoff gullies and rock outcroppings. That was often the job of grunt buddy teams stationed behind a frontline unit to watch for pop-up trouble. So, whatever credit those tanks would get for finally winning the engagement should have been shared with their infantry guides.

> In the words of Captain Frank C. Caldwell, a . . . commander [of F company of 2nd Battalion] in the 26th Marines: "In my view it was the flame tank more than any other supporting arm that won this battle." Tactical demands for the flame tanks never diminished. Late in the battle, as the 5th Marine Division cornered the last Japanese defenders in "The Gorge," the 5th Tank Battalion expended napalm-thickened fuel at the rate of 10,000 gallons per day. The division's final action report stated that the flame tank was "the one weapon that caused the Japs to leave their

Figure 13.1: U.S. Tanks Were Vulnerable to Heavy Caliber Shells
(Source: TM 9-759 [U.S. War Dept., Sept.1944], "Tank, Medium M4A3," p. 8.)

caves and rock crevices and run."[20]

— *Closing in: Marines in the Seizure of Iwo Jima*
History and Museums Division, HQMC

8 March 1945. Despite heavy hostile mortar and small-arms fire, Private First Class Komnenich [of E/2/27] voluntarily led tanks forward of his lines to destroy enemy pillboxes. Advancing ahead of the tanks, he conducted them over difficult terrain and, skillfully directing their fire, rendered valuable assistance in the neutralization of three pillboxes. Later the same day when the tanks were under heavy artillery and small-arms fire and all communications were cut, he braved the hostile barrages to climb on one of the tanks and, while directing fire, sustained a painful wound.[21]

— Navy Cross citation for PFC Louis Komnenich

Some Envelopment Had Been Continually Necessary

Just as the 26th Marines came around the back of the Gorge, so had some envelopment been necessary since the first days of the invasion. Above the Rock Quarry right after landing, U.S. forces had quickly realized the need to outflank any part of a defense line they intended to seize. Despite the examples of Gunny Basilone and others, those in charge would only later realize the need for individuals to also envelop each enemy bunker—through tiny folds in the ground. But, the key word here is "individual." Crawling individuals can do things that a massive hunk of steel can't.

Single Combatant Still More Important Than Any Weapon

Handheld explosives, bazookas, and various kinds of flame weapons would all prove useful on Iwo Jima. But—in the final analysis—it was thinking human beings, and not inanimate weaponry, that would turn the tide. When the preconceived plan no longer worked, the forward-ranging (and unsupervised) American volunteers tried something different.

7 March 1945. Observing three of the enemy setting up

> a machine gun approximately one hundred yards ahead of his platoon's front lines, Corporal Danford [of C/1/26] armed himself with hand grenades and, fearlessly crawling forward to within thirty feet of the enemy's position, hurled his deadly missiles with extreme accuracy, killing two of the Japanese gun crew and driving off the third. "Then, dashing to the gun amidst a hail of enemy rifle fire, he succeeded in dragging the hostile weapon back to his own lines despite a mortal wound sustained en route, thereby preventing the Japanese from regaining the use of the gun.[22]
>
> —Navy Cross citation for Cpl. Cleo S. Danford, Jr.

> 12 March 1945. While delivering a message to the front lines, Private Vargas observed that his company [C/1/27] was receiving intense machinegun fire and suffering heavy casualties from several enemy pillboxes. Courageously exposing himself to the heavy hostile fire, he voluntarily advanced ahead of the front lines and, working his way to one of the pillboxes, silenced it with fragmentation and smoke grenades [probably white-phosphorus to keep the bunker from being reoccupied from below]. Then, proceeding to another one of the enemy emplacements, he neutralized it in the same manner and annihilated the occupants. Although painfully wounded in the chest and left leg, he continued through the withering machinegun fire and, using hand grenades, completely destroyed the third pillbox.[23]
>
> —Navy Cross citation for Private Salvador Vargas

The Marines' Depleted Ranks Had Led to Safer Maneuvers

This book has cited 76 individual contributions that would later lead to medals for bravery. All were chosen for their tactical insights. Of those 76 acts, a full third were from Privates or PFCs. So too had eight of the 22 MoH citations for Iwo gone to E-1's or E-2's.[24] Awards for bravery are least often given to the introductory ranks. That's because—in a top-down bureaucracy—those ranks are considered the least knowledgeable. To make up for this inherent shortfall, their owners are expected to render immediate obedience to all orders. That's the veritable opposite of personal initiative. And most of the nonrates on Iwo had been so far punished for any original idea they had managed to smuggle in from civilian life. Yet, a full third of the tiny forays that ended up saving the mighty divisions (by silencing foe guns and guiding flame tanks) had been by bottom-echelon grunts. If that's not proof of the need for rifleman initiative while fighting a world-class opponent, then nothing is.

In active combat, a Marine rifle company develops a "mind" of its own—one that is only partially influenced by the CO. In effect, the whole becomes more focused than any one part. The CO's role in Vietnam would be to coordinate all unit maneuvers with supporting arms.[25] There, enemy bastions would be attacked the same way as on Iwo. So, a similar role was likely shared by the WWII COs. But, at that tiny Pacific hellhole, the mind of most frontline units had started to reflect the collective opinion of its nonrates. PFCs live in a smaller-scale world than most superiors. It is one of man-sized rocks, and intervening crevices. And, it was only through such a world, that Iwo Jima could be taken with fewer casualties.

Most Long-Range U.S. Fire Wasn't Helping Much

To be truly useful, even the bazookas and flame tanks had to be brought to the front lines. Otherwise, they would have as little effect on well designed fortifications as the standoff artillery and rockets.[26] A direct hit from an aerial bomb would sometimes damage a pillbox, but most were so well camouflaged that they couldn't be accurately marked for the aircraft. Others were on a reverse slope. And still others could be quickly re-manned from below. There must have been some "precision bracketing" by U.S. artillery on Iwo. But, most of its missions had been of the "fire-for-effect" variety—wherein direct hits on pinpoint targets are not common. This 2GW illusion of constantly hurting the opposition would be often repeated throughout the 36-day ordeal.

> The process [final push after a day of rest] began at 7:00 A.M. of March 6, D-Day plus fifteen, in thundering barrages of artillery, naval gunfire, and wave after wave of carrier planes dropping clusters of napalm bombs as their machineguns and rockets ripped into enemy lines in low-

> level passes across the island. In tons of steel and explosives, the firepower approached that unleashed on D-Day in the final hour before Marines hit the beaches.
>
> Virtually every Marine howitzer on Iwo—132 75- and 155-millimeter guns from eleven artillery battalions—was in action. They first fired for thirty-one minutes on the western half of the island, and then rocked the eastern sector for the next thirty-six minutes. In sixty-seven minutes, 22,500 shells slammed down along the front, often as close as a hundred yards in front of the Marines waiting to jump off.
>
> From less than half a mile offshore, a battleship and three cruisers unloaded 450 rounds of eight- and fourteen-inch high explosives. Three destroyers and two landing craft moved close in along the western shoreline to plaster the cliffs and caves with mortars and rockets. . . .
>
> . . . [When the Marine units struck,] Japanese reaction was immediate and deadly across the island. Incredibly, the pulverizing pre-attack bombardment and air strikes seemed not to have affected the enemy at all.[27]
>
> — *Iwo Jima: Legacy of Valor,* by Ross

Most entrances to the big Nipponese blockhouses had been on the backside of elevated ground, making them nearly impervious to direct heavy-caliber fire. To hurt them, a U.S. Navy ship had to sail around behind them. Then, intervening hillocks made the accurate correction of low-trajectory fire difficult. In other words, there had been far too few adequately precise indirect-fire missions on Iwo. But, that was not the only inconsistency associated with 2GW. Many of the pre-assault bombardments had not only created "short rounds," but also warned the foe of an impending ground attack.

> Patrick Couplet, a twenty-five-year-old Fifth Division corporal from New Brunswick, New Jersey, put it more succinctly: "All the bombardment did was let the Nips know we were coming at them again. It stirred them up like a hornet's nest, and the sonsab— were waiting in their caves and bunkers to kill more Marines just as they had every . . . day since we landed."[28]
>
> — *Iwo Jima: Legacy of Valor,* by Ross

But America's traditional style of warfare had called for an impressive display of military might, so that's what TF 56's biggest supporting arms provided. It would be many years, before any high-level U.S. general would publicly admit to surprise being interchangeable with firepower in combat. Mao, of course, had long understood this battlefield paradox. That's why he had said, "It is people, not things, that are decisive in war."[29] But, most Western leaders must have thought he was being too reckless with his soldiers' lives. Only the most insightful seemed to realize what Mao had been talking about.

> Wars may be fought with weapons, but they are won by men.
> It is the spirit of men . . . that gains the victory.[30]
> — Gen. George S. Patton, Jr.

There Would Soon Be a Similar Lesson

U.S. experience has long demonstrated how most safely to penetrate a modern defense line. But, the War Department/Pentagon keeps on claiming the latest technologies are necessary. As at Iwo Jima, they generally help, but are still not the deciding factor. Only light-infantry involvement can limit a defense line's initial degree of effectiveness. The new equipment myth apparently derives from a government agency trying to do too much at once. The job of the Pentagon is to win wars, not to stimulate the American economy through promising, but then not fully testing, more modern equipment.

On Okinawa in May of 1945 (two months after Iwo Jima had been declared secure), 2/4 (formerly the Maoist-oriented 4th Marine Raider Battalion) was called in to do what no other U.S. Army or Marine outfit had been able. It was to seize and hold Half Moon (a.k.a. Crescent) Hill in the Sugar Loaf Defense Complex at the north end of Okinawa's infamous Shuri Line. That Complex was comprised of three hillocks—Sugar Loaf, Horseshoe, and Half Moon. Prior to 2/4's entry into the battle, battalions of the more conventional 22nd and 29th Marine regiments had been trying to seize the Complex for almost two weeks.[31] In that attempt, they had enjoyed all the air, artillery, and tank support they could possibly want. However, by the time those two regiments had permanently occupied only one of the three hillocks on the 18th of May, they had lost almost 4,000 personnel to either battle injury or fatigue.

> During the 10-day period up to and including the capture of Sugar Loaf [Hill], the 6th Marine Division had lost 2,662 killed or wounded; there were also 1,289 cases of combat fatigue. In the 22nd and 29th Marines three battalion commanders and eleven company commanders had been killed or wounded.[32]
>
> — Official Army Chronicle, 2000

That's when, the 6th Marine Division's other regiment was summoned. Within it, lay possibly the Corps' most proficient infantry battalion—2/4. Carlson's 2nd Raider Battalion had been dismantled and its "snoopers and poopers" widely dispersed as machinegunners throughout the Fleet. But, an intact 4th Raider battalion (also Maoist in format) had been redesignated as 2/4. Besides being able to function as a group of tiny elements, the 4th Raiders had experience with an Iwo-like defense belt at Bairoko on New Georgia in July of 1943. While striving to tackle this strongpoint array, it had been deprived of all aircraft, most artillery, and all tank support. Yet, it still managed to get three-fourths of the way through it before dark. From that unexpected success, it had learned an important lesson. Under those kinds of conditions, friendly bombs, shells, and tank treads can be more of a hindrance than a help. Too much headquarters control can also stifle the requisite amount of frontline spontaneity. By all indications, the 4th Raider Battalion had penetrated three of the four rows of mutually supporting bunkers at Bairoko through nothing more than fireteam cooperation.[33]

2/4's initial push against Half Moon on 20 May 1945 did include tanks, but it was also a doctrine-violating double-envelopment. The fire coming from Shuri Ridge in the background then became so intense, that the division commander directed 2/4 to occupy only the forward slope of Half Moon. To accomplish this task, 2/4 specifically requested that there be no tank, air, or artillery support. Its secret weapon would be a ragged row of fireteams in parallel lanes—over which it had relinquished all control. The buddy teams of each fireteam were then to double-envelop the bunker in their lane. But, to do so, they would often have to ask another fireteam to suppress a machinegun firing across the front of their quarry. No squad leader or any other unit leader was to get involved. All coordination would be done between frontline nonrates.[34] So, impromptu improvisation would at times take the place of preconceived procedure. And, the participants' only interference would be from the enemy.

Fast Forward to the Present

For those still having trouble seeing how much effect mid-20th Century events can have on the modern-day world, a more recent example is provided. While most Americans had their eyes glued on Iraq in 2006, there was a telltale event at the border of Southern Lebanon.

> The . . . parallel disasters of Bint Jbiel [sic] and Wadi Saluki became laboratories for teaching how a well-trained insurgent force exhaustively drilled, carefully dug in, camouflaged and armed with the latest precision antitank weaponry could utterly devastate a modern, technologically superior Cold War armored force, even if that force commanded the air absolutely.[35]
>
> – M.Gen. R.H. Scales (Ret.)
> "Armed Forces Journal," December 2007

In effect, *Hezbollah* (and its Chinese and North Korean advisers[36]) had countered a sophisticated "Western" incursion with remotely controlled weaponry that had suddenly emerged from the ground. Because no human operators had been anywhere near the devices for target acquisition and destruction, there were no heat signatures for Israeli aerial reconnaissance to detect.

> [There were unmanned] bunkers housing . . . advanced eavesdropping and surveillance equipment and monitoring cameras.[37]
>
> – *The Times* (U.K.), 10 August 2006

Then, through a marriage of video camera with computer-controlled missile, *Hezbollah* had destroyed over 50 of the world's best tanks (by the Israelis' own count).[38] That was enough of their total to thwart the incursion.

> In one hidden bunker, Israeli soldiers discovered night-vision-camera equipment connected to computers that fed coordinates of targets to the Sagger-2 missile.[39]
>
> — *The Washington Post,* 14 August 2006

Besides Fuel Air Explosive (FAE) missiles and infrared-detonated mines, *Hezbollah* had in place an advanced command-and-control network. Israeli intelligence later found it to contain a "missile-and-rocket" fire control center. So, most of that foreign-supplied ordnance must have been for impeding armor.

The full extent of the tunneling throughout *Hezbollah's* defense line may never be known. Yet, there was an event reminiscent of both Iwo Jima and the United Nations' (U.N.'s) attempt to reenter North Korea in 1953. *Hezbollah* fighters kept reappearing in places that had already come under Israeli control.[40] That made it easy for them to finish off immobilized armor.

> *[Hezbollah]* fighters [kept] constantly popping up . . . , firing, and then vanishing again.[41]
>
> — *The Times* (U.K.), 10 August 2006

All manned emplacements were well camouflaged, ground embedded, and likely connected by tunnel. To locate one, Israeli infantrymen had to be right in front of it.[42] That will sound all too familiar to the veterans of Iwo Jima, Korea, and Vietnam.

Through a combination of video camera, computerized missile, and guerrilla-like spoiling attacks, *Hezbollah* had faced down 400 sophisticated battle tanks and their equally advanced air cover. The Israelis' subsequent end run (traversing the line from a flank) then made no difference to the final outcome.[43]

Of course, a few *Merkavas* were killed the old-fashioned way. During Israel's long occupation of Southern Lebanon, *Hezbollah* had learned how remotely to detonate a sub-surface charge where Israeli tanks became channelized. That charge did not have to be "shaped," only big enough to lift the entire vehicle off the ground. When it came back down, the crew was dead from concussion.

Would Tiny 3GW Forays Work at Today's 38th Parallel?

Long before remotely controlled tank killing, deliberate defenses were on occasion so well-meshed with the surrounding terrain that they became impregnable. Such was the case on the western end of Okinawa's Shuri Line in May of 1945. After assuming responsibility for that sector from the Tenth U.S. Army and fighting for 10 more days, the 6th Marine Division commander would finally decide to take only the front face of Half Moon. His other units were then able to complete the attack on Horseshoe (the other Sugar Loaf holdout) *en route* to the Asato River and Naha City.[44]

Some eight years later (in 1953) during the Korean War, the North Koreans and their Chinese allies would employ a fire plan so carefully tailored to the hilly landscape that it wood keep the mighty U.N. phalanx from breaking through their 38th-Parallel defense belt.

But, lightly armed infantrymen are able to encroach upon an MLR of this magnitude, whereas heavily supported conventional forces cannot. Skilled infiltrators can even stop an armored counterattack. One of their favorite methods over the years has been stationing one man below ground right next to a narrow tank trail. When no opposition soldiers are looking, he drags a landmine—by string or wire—under the tread of approaching armor. That trick has been reported in several widely separated parts of the world: (1) the Finns in 1939;[45] (2) the Japanese during WWII;[46] and (3) the Viet Cong (VC) in Vietnam.

The advantage of authority-delegated light infantrymen over headquarters-controlled heavy infantrymen, is—after all—the increased agility and initiative of a tiny maneuver element. And, fewer lives will be ultimately risked that way. Expert light infantrymen need few, if any, supporting arms to accomplish any mission. This, in turn, eliminates the possibility of a large friendly fire accident. A few 3GW-like infiltrations of a modern-day defense belt therefore become the preferred alternative to a concentrated 2GW onslaught.

What Makes This Still True

One might not be able to completely overwhelm a defense line with infiltration teams. But, by disabling some of its anti-armor and anti-personnel systems, a few teams could pave the way for a more conventional push.

All defense belts attempt to mesh fire plan with real estate. But, few of God's creations are totally uniform. So, as at Iwo and Okinawa, every objective's natural surroundings will offer a few gaps. If one can't be found at the customary scale, then crawling "recon pulls" must look for one at a smaller scale.

The most sophisticated overhead surveillance systems still can't see through any obstruction—whether it be solid like a roof or proliferated like foliage. Even out in the open, they will have trouble distinguishing well-camouflaged men from hot rocks at sundown, and well-clothed soldiers from rabbits later at night. So, no amount of modern-day paraphernalia will prevent all short-range infiltration.

Most ground battles can only be decisively won at the infantry level. Despite all well-intended efforts to fight them at a distance, that in no way guarantees a successful occupation. Thus, any new infantry gear can have the effect of an unnecessary "crutch"—by

Figure 13.2: What May Be a Hdqts. Element Coming Ashore
(Source: Wikimedia Commons, image superimposed on back of photograph entitled "Iwo Jima Cemetery, 1945," Ralph J. Vaccaro Collection at USMC archives, ©.)

limiting the amount of frontline maneuverability and spontaneity. For example, no radio encryption device will ever be as useful as a Navajo codetalker.

Moving Forward from Here

Despite America's failure to decisively win any extended ground conflict since WWII, its overall culture remains decidedly Western—"top-down" in how most problems are solved. Meanwhile, the Asian Communist regimes—though highly authoritarian and less interested in human life—allow their armies to operate from the "bottom-up" (in concert with Eastern culture). So, where strategy has been driving operations and operations dictating tactics for U.S. units, the opposite is true with their Eastern counterparts. (See Figure 13.2.) For those traditional foes, successful tactics have been shaping operations, and then operations forming strategies. This has not been the result of developing-nation backwardness, but rather universally acknowledged common sense. It's how a ragtag group of Colonial Minutemen kept a better-armed British force from marching on Concord in 1775.

In fact, under most wartime circumstances, this reverse Eastern progression is more effective than the Western hierarchy, and military historians know it. But, because new military equipment helps to stimulate the U.S. economy, Washington has been permitting an ongoing lack of tactical emphasis. At Iwo Jima, this minimal portfolio of small-unit maneuvers was temporarily augmented through frontline improvisation. But since that time, too much headquarters control has infected all ground branches of the U.S. military. That's why it is so important for all service members to remember the "Iwo Alamo" before motoring off to another big war.

14 If Iwo Jima Were Stormed Today

- What had been the most useful weapon on Iwo?
- Would that weapon now obey the Geneva Conventions?

Modern-day artillery is bigger and more accurate.

(Source: FM 22-100 [1973], p. 2-11.)

Overall Impression of the 1945 Battle

Those who have had to fight for their own survival can best appreciate what happened on Iwo Jima. War is one of man's most obscene inventions. Some say that justifies any excess in its execution. But, the Marines who had landed on Iwo were mostly Christian and well-disciplined. They knew that wherever God's children must go, He would be there with them. Whatever degree of conflagration they would have to endure, He would endure as well.

"Look!" he answered, "I see four men loose, walking in the

midst of the fire; and they are not hurt, and the form of the fourth is like the Son of God."[1]
— Daniel 3:25, *New King James Bible*

The assault on Iwo so quickly morphed into a nightmare situation, that the *bravado* young men sometimes display in the face of danger soon vanished. Then, there were two kinds of people around: (1) those who were trying to avoid enemy fire by digging in; and (2) those who were trying to avoid it by moving. The generally accepted definition of courage is continuing to function while terribly afraid. So, for every Marine who later received a medal for bravery on that tiny island, there were hundreds who should have. But, the combat was so intense and long-lasting that most of their exploits would be witnessed by soon-to-be-injured buddies or simply lost in the shuffle of after-action reports.

It would be the height of hypocrisy to now try to second-guess any American who fought for Iwo Jima in 1945, from the most senior general down to the newest Private. Their selfless endeavors have been forever emblazened on the walls of honor and justice. They, to a man, had done their absolute best under very difficult conditions. In fact, the frontline troops had done more than is even imaginable after what has since been discovered about enemy capabilities. Only the Holy Spirit within those young GIs could have inspired such a super-human effort. It has been generally acknowledged that they fought more for each other, than any other reason. They had shared a bond of brotherhood forged by the Corps, and didn't want to let each other down. But, the degree of love necessary to give up all that life has to offer for another human being transcends all worldly parameters. Those exemplary young Americans had continued to risk almost certain death, because of the faith they had that they and their buddies would somehow survive.

With Regard to Other Issues

This book has hopefully demonstrated that unit cohesion had more to do with the Marines' victory on Iwo than any technological marvel. That cohesion was even able—through the concurrent tradition of assuming a superior's responsibilities—to compensate for a tactical mismatch. Of course, flame weapons, bazookas, and demolitions would also prove useful. But, most of America's stand-off firepower had failed to meet expectations. And, according to some division staffs, there had been too little teamwork within recently augmented fireteams.

> The new men, expected to replace invaluable veterans of the Pacific War, were not only new to combat, but they also were new to each other, an assortment of strangers lacking the life-saving bonds of unit integrity. "They get killed the day they go into battle," said one division personnel officer in frustration. Replacement losses within the first 48 hours of combat were, in fact, appalling. Those who survived, who [quickly enough] learned the ropes and established a bond with the veterans, contributed significantly to the winning of the battle.[2]
>
> — *Closing In: Marines in the Seizure of Iwo Jima*
> History and Museums Division, HQMC

But, had the new guys fared poorly or simply encountered more separate-fireteam action in the narrow confines of the northern plateau. Technique familiarity is important in heavy combat, but so too is a willingness to try new things. Beauty is in the eye of the beholder. The recent joinees would have been perfect for a nontraditional (3GW-like) mission. And, all young grunts of this era were like brothers. Four who had just met could have—within hours—established enough chemistry to fulfill the fireteam role within the most common (2GW) formations. Many of those replacements would end up winning bravery medals.

Not until about 1973 would U.S. infantry cohesion really start to deteriorate. Largely to blame has been the mandatory adoption of new societal norms. Possibly to help with the effect on discipline, small-unit tactics have become more standardized. And, flame weapons are no longer encouraged by the Geneva Conventions. So, whether or not 40,000 modern-day U.S. infantrymen could capture that miserable little pile of rock from 20,000 WWII-vintage Japanese is very much in doubt.

The All-Important "Togetherness" Factor

Two events would lower the chance of contemporary Marine

rifle companies developing enough unity: (1) the disempowerment of each company's top enlisted leaders around 1973; and the removal of squad (or platoon) "sleeping" bays in the 1980's.[3]

When all personnel lived side-by-side in the same huge room, and their enlisted leader in an adjoining cubicle, something very special happened. The SNCOs whose job it was to advise COs on all personnel matters had little trouble keeping up with troop needs, opinions, and insights. Then, the two-man-room came along. With it went away the most important part of that upwards communication, as well as the family structure upon which good infantry outfits depend.

Prior to 1973, each company's First Sergeant had handled all disciplinary problems. Then, with administrative control shifting to battalion and all nonjudicial punishment suspended, unit morale was not the only thing to suffer. Every Leatherneck's "minor-error history" had previously been expunged from his file every time he got transferred. Now, only those with near-perfect records were able to reenlist. Most had been careful to do nothing controversial. But, judgmental errors can be indicative of valuable initiative. And, mistakes corrected in peacetime will not generally reoccur in combat.

Before 1973, the infantry company Gunny Sergeant also took care of all enlisted training, to include that which was tactical in nature.[4] Then, he became nothing more than a "bean counter"—the monitor of unit resupply. Regardless of his personal level of combat experience (often many years of overseas deployment), he was no longer allowed anywhere near the maneuver instruction. Recent college graduates would instead control all enlisted instruction on every subject.

The GIs' Eroding Affiliation with Organized Religion

In 1945, religion was considered an important part of every Marine's welfare. Just prior to the Iwo landing, here how all participants were prepared for their ordeal.

> Everything had been done for them [the assault troops] that an efficient military organization could do—their equipment was probably the best in the world; they had been well fed;

> they had attended religious services; they had been intelligently and thoroughly briefed.[5]
>
> —*WWII History of 5th Marine Division,* by Connor

Today, God is no longer allowed as much influence on American society. His teachings have been banned from U.S. schools and courthouses. In a contemporary Iwo-like struggle, that could have an effect on GI perseverance.

> Fortitude is . . . a gift of the Holy Spirit.[6]
>
> — "Encyclopedic Dictionary," *The New American Bible*

Tactical Flexibility Still Important

In the late 1940's, the Marines' mission was to close with the enemy, not overcome him from a distance with some advanced piece of equipment. As the first few chapters have shown, the defenders of Iwo Jima were employing state-of-the-art (rearward-moving and soft) strongpoint matrices. Fortunately, the American ground forces were eventually able—through frontline improvisation—to breach those advanced defensive arrays.

3/9 had broken through the Third Strongpoint Belt with a bombardment-free nighttime infiltration.[7] And, other units had made huge strides in how each bunker could be more safely approached in broad daylight. It took an envelopment through folds in the ground and sometimes a pole charge. Unfortunately, little of that tactical innovation would make it into their organizational knowledge after the war. Most Leatherneck infantrymen no longer practice either microterrain appreciation or short-range infiltration.

Firepower Now Has a Viable Substitute

There's no question that firepower played a role in Iwo's capture, but the most productive kinds posed little risk to their users. So during the final infantry assault, "fire superiority over the enemy" or "keeping defender heads down" had been eventually replaced with "surprising bunker occupants." The most useful ordnance on Iwo then became flame tank, bazooka, and hand-delivered explosive.

The Flame Weapons

Four flame weapons would be used on Iwo: (1) napalm bombs; (2) man-carried flamethrowers; (3) "zippo" tanks; and (4) the "corkscrew and blowtorch" method (igniting diesel fuel just poured down some hole in the earth). While the enemy's underground structures had been built to withstand such an attack, their surface portals were still susceptible to it.

After the fire-bombing of Tokyo and a few other heavily populated areas during WWII, all flame weapons became eventually discouraged by international agreements.

Flamethrowers have not been part of the U.S. inventory since 1978, when the Department of Defense (DoD) stopped using them. Deemed of questionable value in modern combat, they also presented a public-relations problem (through the unpleasant death they inflicted). Yet, their wartime use has only been completely banned from military targets within a civilian-occupied area. Most military use of flamethrowers was stopped by the Convention on Certain Conventional Weapons (concluded on 10 October 1980 and then enforced in December 1983). Washington did not ratify the Prohibitions or Restrictions on the Use of Incendiary Weapons (Protocol III), until 23 December 2008.[8]

Other Standoff Weaponry

Of late, the Pentagon has become quite proud of how easily it can pinpoint an enemy target from the air. A Youtube video of January 2020 shows a U.S. Air Force drone operator in Nevada purportedly destroying a small Islamic State in Iraq and Syria (ISIS) convoy on the night of 1 June 2019. To do so, he used a Predator-mounted Vulcan gatling gun. Through thermal imaging, he could target not only the vehicles, but also everyone trying to escape them on foot. The moral rectitude of so thorough an execution aside, most viewers were left wondering how the convoy had been positively identified as ISIS.[9]

Such a weapon system would have had little effect on Iwo. Before and after its invasion, the enemy had moved around almost entirely below ground. After seeing so little activity, the U.S. Navy would shorten its planned bombardment by a full week.[10] While many short-range infiltrators were later spotted at night, few defenders would openly advertise their exact whereabouts during the day.

> Seldom on Iwo, from D-day until the battle was over [36 days later], did you see the enemy—just the sights and sounds of deadly fire from his weapons.[11]
>
> — Marine infantryman on Iwo Jima

Drone and Satellite Reconnaissance

Today's surveillance devices would be of no more help on Iwo than aerial photography. After the Marines had landed, the Japanese rarely came to the surface, and only then to run tiny nighttime raids. Enemy troops were only occasionally seen moving around above ground in significant numbers. This same lack of observability also applied to their protective arrays. Almost all surface emplacements had been carefully camouflaged, to include trenchlines and indirect-fire gun pits. (See Figure 14.1.) Even the spider holes were fitted with trapdoors, and pillbox slits with metal covers.[12] This unexpected degree of obscurity has been well documented.

> Marine spotters in Maytag Messerschmitts watched the attack come to life in brilliant, chilly sunlight. Theirs was a familiar spectacle on Iwo, but totally foreign to anything else in military history. Below them was a battlefield where one army fought above ground and the other fought almost totally beneath it; where thousands of troops moved in the same area at the same time, Marines maneuvering on the surface in the attack, the Japanese moving in tunnels from one underground strongpoint to another whenever the tide of combat called for more firepower or reinforcements.
>
> Everything beyond the line was completely barren and devoid of any sign of enemy troops—a rugged, uninhabited landscape as mysterious as a solar planet.[13]
>
> — *Iwo Jima: Legacy of Valor,* by Ross

The island, merely eight-square miles, was riddled with a sophisticated system of enemy tunnels. Moving from one objective to another, Marines were easy targets for the

Figure 14.1: Even Mortar Pits Were Invisible from the Air
(Source: "Japan's Battle of Okinawa April-June 1945," by Thomas M. Hubler, Combat Studies Inst.,U.S.Army Cmd.& Gen.Staff College, p. 62.)

> Japanese who were able [to] duck into this "basement of catacombs" and maneuver behind the American troops, shooting them in the backs.[14]
>
> — *Camp Lejeune Globe,* 8 October 1999

> From the start of 1945, the digging of the underground passageways known as *dokutsushiki kotsuro* (cave-style communication tunnels) began in earnest. Once they were completed, they would enable [Japanese] soldiers to move from one installation to another without having to emerge above ground.[15]
>
> — *So Sad to Fall in Battle,* by Kakehashi

Before long, the Marines started running into the same people they had just fought on the last ridgeline. And those people seemed as well armed as before. That's also the beauty of a state-of-the-art urban defense—all assets channeled to where the opposition is actually attacking, and then unexpended "portions" shifted rearward.[16]

> By the time we were moving to the north, probably around the middle of March, we began to realize they [the Japanese] were fooling around underneath us. . . . What Gen. Kuribayachi had done was arrange a series of defensive positions all the way up in the north. Once we would kill enough in one, they would move back to the next one, either at night or by tunnel.[17]
>
> — CO of E/2/25 in a later interview

Most New Gear for the Better Targeting of Enemy Personnel

Much of the latest U.S. gear has to do with more efficiently destroying enemy soldiers. But they can't be killed until they are precisely located. The latest surveillance equipment cannot yet detect those who are fully dug in with overhead cover. In fact, it has trouble identifying any heat signature through heavy vegetation. So all adversaries with the ability to maneuver below ground will be nearly immune to U.S. firepower.

Some American devices now theoretically detect man-made structures below ground. But, they have limited utility in active combat.

Ground Penetration Radar

A modern-day American technocrat or arms manufacturer might claim the entirety of Iwo Jima's underground labyrinth would now be visible through Ground Penetration Radar (GPR). The latest GPR devices can—to some extent—locate tunnels, bunkers, and caches. But, this is from ground level (not some moving aerial platform), and much depends on a uniform composition of overhead medium (soil conditions). Within rocky ground, for example, it produces indistinct images. And the landscape easiest to excavate is often the east receptive to high-frequency radio waves.[18]

Such GPRs are too heavy to be man-carried. They are also time-consuming to use and vulnerable to enemy fire. As a result, they are more easily applied to rear-area road clearing than front-line combat. On Iwo, they would have had limited utility.

Other Enemy Detection Devices

To identify danger, GIs have been equipped by their Creator with six senses (with the sixth arguably being a subliminal combination of the others). Any intruder detection device that enhances one sense may detract from others. That's why the last invention to significantly alter ground combat has been the machinegun. Expert light infantrymen need few technologies to rule the battlefield. They view the latest gadgetry as an impediment to their former agility/maneuverability.

Image Intensification Devices (Night Vision Goggles)

All Night Vision Goggles (NVGs) require some ambient light to operate. There is no such thing in a tunnel. Nor can the latest NVGs see well through bushes, smoke, dust, haze, or adverse weather conditions.[19] One of the best ways to defeat them during a night attack is to run a mortar illumination mission a mile or so behind the assault force.

Thermal Imaging (Infrared Sensor) Devices

Thermal imaging cannot distinguish one heat signature from another. Iwo Jima was covered with volcanic activity. There had been a myriad of natural heat signatures there. In trying to dig a tunnel between Airfield No. 1 and Suribachi, the Japanese had run into high temperatures. Around Motoyama Village, Kuribayashi's letters talk about ground that was 140° Fahrenheit and spurting sulphur vapors. That's where one of his headquarters was located about halfway between the Second and Third Defense Belts. In fact, Motoyama Village sat atop Iwo's other volcanic cone. So, both defense belts may have contained hot ground. 200 yards west of Motoyama Village lay a sizeable sulphur mine, and there may have been another just northeast of the village as well.[20]

> At the southern end of the island and around the airfields, Marines still shivered in their foxholes at night. But farther north they had come to the Japanese sulphur wells.[21]
> — *Battle for Iwo Jima, by* Leckie

> Marines could scarcely dig a foxhole without starting a sulphur bath. They could cook a can of C rations by burying it in the earth for a quarter-hour. When they wanted to make coffee, they took their canteens to the sulphur wells. Sometimes the temperature of the water rising to the surface was 160 or 170 degrees.[22]
>
> — *Battle for Iwo Jima, by* Leckie

Most of the Japanese troops fighting for Iwo had overhead cover that would have blocked their own heat signature from aerial detection. Iwo was a tiny speck in the middle of a very big pond. It had rained there during the first three days of the invasion. Thermal imaging devices have trouble seeing through water vapor and precipitation. To them, individual droplets or flakes look black. In dense fog, they will not work at all. An "ocean haze" was recently enough to completely obscure a drug boat being chased by a thermal-imaging-equipped U.S. Coast Guard helicopter north of Cuba.[23]

Anti-Personnel Radar

In what is fast becoming a virtual-reality battlefield, another type of early warning device has long intrigued armchair warriors. To detect an unseen enemy, it first took the form of remotely controlled ground sensors in Vietnam. Then, it became like the motion detectors for home security. Finally, it—for a while—purported to help U.S. small-unit leaders determine which obscured nearby areas were actually occupied. But, "radar flashlights" would have little luck through the thick concrete walls or tunnel bends of an Iwo setting.

> . . . The RADAR Flashlight uses a microwave radar combined with a very fast signal processor to detect minute movement of the human thorax due to the person's heartbeat or breathing. . . .
>
> The RADAR Flashlight can detect respiration of a human behind a wall, door, or in an enclosed space with nonconductive walls at distances of up to about 10 to 16 feet, depending upon the material. It can also detect a person behind foliage, such as bushes, amongst a tree line, or be-

> hind a chain link fence. It can detect an individual behind a solid wooden door or standing 4 feet behind an 8-inch block wall.[24]
>
> — *Marine Corps Gazette,* January 1999

The Greater Firepower of a Modern-Day Invasion Phalanx

Today's towed artillery is far more destructive than that of 1945. But, even the pillboxes and caves successfully damaged by big bombs and shells on Iwo had been quickly remanned. So rubbling a battlefield helps little with its capture.

Thermobaric Munitions

Thermobaric munitions (sometimes called FAEs) are flame oriented and thus sometimes in conflict with the Geneva Conventions.

> "What kills [personnel] is the pressure wave, and . . . subsequent . . . [vacuum], which ruptures the lungs. . . . If the fuel deflagrates but does not detonate, victims will be severely burned and will probably also inhale the burning fuel. Since the most common FAE fuels, ethylene oxide and propylene oxide, are highly toxic, undetonated FAE should prove as lethal to personnel caught within the cloud as most chemical agents."[25]
>
> — *Forbes Magazine,* March 2017

Kuribayashi's ground level fortifications had been specifically designed (and then intentionally built of high-quality and steel-reinforced cement) to counteract the effects of U.S. concussion and flame.

> [T]he way he [the Japanese general] had built and placed his fortifications, made shelling and bombing largely useless. The positions were masterpieces of concealment and construction: walls of many were more than three feet of steel-reinforced concrete and impossible to spot from the natural terrain upheavals that camouflaged them. Some

> measured ten by twenty feet, had five-foot ceilings, and were made up of three separate rooms with twelve-inch concrete walls, the areas connected by narrow crawl-ways.
>
> Apertures for cannon were virtually invisible, jutting from rubble of [the] volcanic boulders. Atop many of the installations were pillboxes . . . all but impregnable fortresses that could take a direct hit from naval shell fire or a bomb . . . without suffering much damage. Even when Marine demolitions men could fling a satchel charge into a gun vent, the blast would have little or no effect on the other two rooms.[26]
>
> — *Iwo Jima: Legacy of Valor,* by Ross

Most sub-surface passageways on tiny Iwo had more than one level and many exits. (Look back at Figures 2.5 and 2.9.) They were thus well protected from poison gas, concussion, "corkscrew and blowtorch" technique, and all surface-induced vacuums.

> [T]unnels would veer off at right angles halfway down their length to neutralize bomb blasts, while entrances and exists were built at different levels to improve ventilation.[27]
>
> — *So Sad to Fall in Battle,* by Kakehashi

America's New Bunker-Buster Munitions

A thermobaric GBU-43/B Massive Ordnance Air Blast (MOAB) bomb was dropped over Islamic State positions in Afghanistan's eastern Nangarhar Province in April 2017. It had been intended—through the depletion of all available oxygen—to discourage the Taliban's use of cross-border tunnels in the region. However, this first use of a MOAB in wartime only proved encouraging because it had been dropped in a narrow canyon filled with cavern entrances.

> The Air Force statement said that the weapon is designed to clear targets in caves and canyons (about the giant bomb).[28]
>
> — *Bloomberg News,* April 2017
>
> The canyon walls had magnified the bomb's effect on all em-

bedded caves. Had the same bomb been deposited atop a fortified mountain or open plain, it would have been much less successful. Then, most of the blast would have been dissipated down hillside slopes or across unobstructed space, and anyone deep beneath spared from harm. Such would have been the average target on Iwo.

Of course, DoD has also been developing a Massive Ordinance Penetrator (MOP) to succeed the BLU-118/B "Bunker-Buster" of the early 2000's (which could go through over six feet of concrete before exploding).[29] Yet, even this newest MOP would have a minimal effect on pillboxes that could not be precisely located from above.

> The latest GBU-28 laser-guided bunker buster can [downward] penetrate [up] to 30m [of rock].[30]
>
> — Director, Center for Korean Affairs (Seoul)

Few of Iwo Jima's tunnels were deeper than 30 meters, but their permanent closure in the easy-to-dig and shock-resistant stone would have taken the direct hit of an oversized munition.

The Modern-Day Equivalent to an Iwo Nightmare

As in the Western Pacific of WWII, underground fighting need not be restricted to defensive missions. It can be easily carried into the offensive realm—as with spoiling attacks from tunnels that now reach back to overrun strongpoints. Should North Korea again invade its democratic neighbor, the first evidence may be many miles to the south of the current Demilitarized Zone (DMZ). That's where exits to so-far-undiscovered passageways would be opened. Some opposition units might even emerge from Seoul's subway system.[31]

While the U.S. military does not practice much below-ground combat, many of its past adversaries have. They are most notably the North Vietnamese Army (NVA) in Vietnam and Taliban in Afghanistan. Decisively defeating either one would have taken invading their homeland. That's where an Iwo-like zone might have been encountered.

North Korea is the most dangerous of America's present-day adversaries. Everything between Pyongyang and the frontier has been extensively tunneled and then filled with huge, nuclear-bomb-resistant chambers.[32] Unlike the below-ground network on Iwo, North Korea's much bigger sub-surface system could not be restricted from outside resupply and reinforcement.

If Napalm-Spreading Supporting-Arms Were Still Available

America quit using napalm bombs around 2003. And the last U.S. flame tank—the M67—went out of active service in 1974.[33] But, even with the latest additions to the Geneva Conventions,[34] any successor to the M67 would have little legality at a modern-day Iwo Jima. That's because there had been up to a thousand forced Korean laborers on that tiny island during the U.S. invasion.[35] Even recently drafted local inhabitants (who had also existed on Iwo) would today constitute civilian hostages.

DoD Must First Admit to the 1945 Accomplishment

The average Marine who fell on Iwo Jima was 22 years old.[36] Many frontline riflemen were much younger. Without trying to imagine their unfilled hopes and dreams, one cannot hope to fathom the depth of their sacrifice. That sacrifice went far beyond the call of normal duty. When killed, many were personally trying to save their parent-units from the ravages of a sophisticated MLR. For, it was in their era that the current state of the art for a rural defense (and how safely to penetrate one) first appeared within a number of major armies. But, all were from the Eastern World, because their Western counterparts had been focusing too heavily on their own firepower.[37] Iwo had provided a brief "window" into what was possible for U.S. warfighters with as much evolutionary insight. But, after the Okinawa crucible of May 1945, that window would close.

Martial necessities are sometimes hard to recognize amongst economic and political pressures. Continually mastering the latest equipment makes it easy to forgo tactical rehearsals. And, getting promoted can seem more important than the welfare of subordinates. But those in charge of future U.S. grunts will continue to experience guilt if they don't start better preparing their people for 3GW combat. That they have done everything directed will provide little solace. It is their sacred duty to weigh all competing factors and then do the right thing. The following prayer may be of help in navigating what the Catholic Church calls "blind spots" in every person's outlook. It was said by Sgt. Joe Enders as he lay dying on Saipan or Tinian in the movie "Windtalkers."[38] To resolve competing responsibilities, he had managed to accomplish the mission while still allowing his men enough tactical leeway to save themselves from destruction. Good leadership does, after all, require some sublimation to the whole.

HAIL MARY, FULL OF GRACE. THE LORD IS WITH THEE. BLESSED ARE THOU AMONGST WOMEN, AND BLESSED IS THE FRUIT OF THY WOMB, JESUS. HOLY MARY, MOTHER OF GOD, PRAY FOR US SINNERS NOW, AND AT THE HOUR OF OUR DEATH. AMEN.

Breaching a Modern Defense Belt Takes Infantry Infiltration

After the Battle of Iwo Jima, the role of the American infantry officer should have changed. The sub-surface nature of the battlefield and sapper ability of the enemy had created consolidation problems after every U.S. advance. In effect, all defensive formations became porous. Their rear areas could be only secured through a blanket of pro-active sentry posts. Such assignments had been so numerous on Iwo as to be mostly filled by nonrates. Low-ranking volunteers had also made contested large-unit maneuvers less dangerous—through enemy machinegun silencing and friendly tank guiding.

So, the primary role of the combat commander was no longer one of enforcing established procedure or precisely following headquarters' instructions. It was now one of facilitating the collective effort of rank-and-file unit members.

After the war, Gen. Erskine would become the USMC's technical adviser to the "Sands of Iwo Jima." That movie would then focus on the Marine rifle squad and its individual members. Not until the insights and improvisations of frontline U.S. riflemen are again appreciated, will America win another extended ground conflict against a 3GW-proficient opponent.

NEVER FORGET ALL THOSE FINE YOUNG AMERICANS ON IWO JIMA.

Notes

SOURCE NOTES

Illustrations:

The image on the cover and page 3 is from "Fortitudine," volume 24, number 4, Spring 1993, page 5. This is a bulletin of the Historical Division, HEADQUARTERS MARINE CORPS. As "Fortitudine" is a government periodical, all contents are supposedly in the public domain. This map of Iwo Jima was probably done on active duty by a veteran of the Battle for Iwo Jima—Col. Charles A. Waterhouse. Unfortunately, he can no longer be contacted to make sure. So, as an educational nonprofit, Posterity Press considers the reprinting of this low resolution version of his work to be fair usage. It was retrieved in December 2019 from the following website address: https://www.marines.mil/Portals/1/Publications/Fortitudine%20Vol%2024%20No%204_1.pdf. Others must be more mindful of a copyright. All rights reserved. Copyright © n.d.

Pictures on pages xviii ("Flotsam and Jetsam"), 62 ("Fire in the Hole"), 133 ("The Grenade"), 171 ("The Gorge"), and 195 ("The Target") are by Charles H. Waterhouse. They come from the selections in *Closing In: Marines in the Seizure of Iwo Jima,* by Col. Joseph H. Alexander from the Marine Corps Combat Art Collection. As part of the Marines in World War II Commemorative Series pamphlet, from HEADQUARTERS MARINE CORPS in 1994, these low-resolution images are now in the public domain.

Picture on page 6 reprinted from Posters Section of U.S. ARMY'S CENTER OF MILITARY HISTORY. As it appears to have been painted by an active-duty member, it is in the public domain with image designator, "1-25-49.jpg."

Picture on page 23 reproduced after telephonic assurance from Osprey Publishing Ltd., London, that this is a U.S. War Dept. (public domain) sketch in *JAPANESE ARMY OF WORLD WAR II,* Men-at-Arms Series, text by Philip Warner, color plates by Michael Youens, page 25. Copyright © 1972 by Osprey Publishing Ltd. All rights reserved.

Picture on page 24 has been reprinted after extraction from the videocassette documentary "A Tribute to WWII Combat Cameramen of Japan," from NIPPON TV, n.d. This low-resolution application for educational purposes is considered fair usage. Others must be more mindful of a copyright. This same picture most recently appeared in "Reflections on Iwo Jima: The Japanese Tunnels," Youtube presentation from *Marines TV,* on 26 March 2011, at the following url: https://www.youtube.com/watch?v=NVkrhvqlyOY.

Picture on page 31 has also been reprinted after extraction from the videocassette documentary "A Tribute to WWII Combat Cameramen of Japan," from NIPPON TV. This enhancement to a low-resolution version for educational purposes is considered fair usage. Others must be more mindful of an original copyright. Its details were checked against a somewhat more visible rendition in "Reflections on Iwo Jima: The Japanese Tunnels," a Youtube presentation from *Marines TV,* on 26 March 2011, at the following url: https://www.youtube.com/watch?v= NVkrhvqlyOY.

Picture on page 32 has been reprinted with permission of HOWARD MODELS, in Toledo, OH. With image designator "IwoJima2s.jpg," this low resolution painting was retrieved in December 2019 from the following url: http://www.howardmodels.com/0-topographic/terrain-models/Iwo-Jima/Iwo-Jima-2.html. All rights reserved. Copyright © Howard Models, n.d.

Map on page 52 reprinted from HISTORYOFWAR.ORG. Its image designator is "IwoJima8m.jpg." For a nonprofit educational publisher, its reproduction is deemed fair usage. Others must be more careful of a possible copyright. All rights reserved. Copyright © historyofwar.org, n.d.

The picture on page 57 is from "First Offensive: The Marine Campaign for Guadalcanal, by Henry I. Shaw, Jr. Drawn by Donald L. Dickson. it comes from page 25 of the Marines in WWII Commemorative Series pamphlet by HEADQUARTERS MARINE CORPS of 1992. It was retrieved in October 2018 from the following website address: https://www.marines.mil/Portals/59/Publications/First%20Offensive-The%20Marine%20Campaign%20for%20Guadalcanal%20%20PCN%2019000311700_3.PDF. All contents of a government publication are theoretically in the public domain. So, as an educational nonprofit, Posterity Press considers the reprinting of this low resolution black and white version of the original artwork to be fair usage.

Picture on pages 74 and 114 are in the public domain from WIKIMEDIA COMMONS. They have image designators "800px-Seventh-War-Loan-Drive-Poster-1945.jpg" and "Military Transport Boats at Iwo Jima.png."

Pictures on pages 79 and 97 reproduced after written assurance from Cassell PLC, London, that the copyright holder for *WORLD ARMY UNIFORMS SINCE 1939,* text by Andrew Mollo and Digby Smith, color plates by Malcolm McGregor and Michael Chappell, could no longer be contacted. The illustrations are from part I (plate 155), and part II (page 71) of the Cassell publication. All rights reserved. Copyright © 1975, 1980, 1981, 1983 by Blandford Press Ltd.

Picture on page 109 reprinted from Artphoto Archives of the U.S. ARMY'S CENTER OF MILITARY HISTORY. Appearing to have been painted by an active-duty member, this picture is considered to be in the public domain. It has the following image designator: *0806-3.jpg."

Picture on page 116 is from "Marine Corps History," volume 1, number 1, of Summer 2017, page 30. From the History Division, HEADQUARTERS MARINE CORPS, "Marine Corps History" is a government periodical with all contents supposedly in the public domain. This image was probably done by an active-duty U.S. Army member. But even if it wasn't, as an educational nonprofit, Posterity Press considers the reprinting of a low-resolution version to be fair usage. It was retrieved in May 2020 from the following website address: https://www.usmcu.edu/Portals/218/MCH%20Vol%203%20No%201.pdf?ver=2018-10-16-115407-260. Others must be more mindful of a copyright.

Picture on page 121 entitled "Us Marines" is a painting by Charles Waterhouse. It comes from the August 1985 cover of *LEATHERNECK MAGAZINE*. In 1971, Waterhouse received a commission as major from the Marine Corps History Division. At some point thereafter, he became the "USMC Artist in Residence." He served in that post until his death as a Colonel in 1991. This use of a low-resolution screen shot of that cover by an educational nonprofit is considered fair usage. But, others must be more mindful of a copyright.

Map on page 138 and 139 is from what is currently a BALLANTINE BOOKS title, *So Sad to Fall in Battle: An Account of War Based on General Tadamichi,* by Kumiko Kakehashi. While this title was first published by Presidio Press in America in 2009, it was originally from Japan. As an educational nonprofit, Posterity Press therefore considers the reprinting of this low resolution map to be fair usage. Others must be more mindful of a copyright. Parts of the book were retrieved in March 2020 from the following url: https://books.google.com/books?id=GTUVD9qN7dQC&printsec=copyright#v=onepage&q&f=false.

Picture on page 230 has been reused in accordance with a license from CREATIVE COMMONS. An apparent painting superimposed over a photograph entitled "Iwo Jima Cemetery, 1945," this is an official USMC photograph in the Ralph J. Vaccaro Collection (COLL/5245) at the Marine Corps Archives and Special Collections that may be partially copyrighted ©. But, the reprinting of this poor quality image of it by an educational nonprofit is considered fair usage.

ENDNOTES

Page vi:

1. "Obituaries: Colonel Charles Waterhouse USMCR," *The Star-Ledger* (Newark, NJ), 18 November 2013; "USMC Museum Receives Col. Charles H. Waterhouse Art Collection," Marine Corps League, *Semper Fi Magazine,* vol. 68, no. 5, September/October 2012, pp. 18, 20; both in *Wikipedia Encyclopedia,* s.v. "Charles Waterhouse (artist)."
2. Third Marine Division Association. *CALTRAP Magazine,* vol. 52, no. 2, March/April 2008, pp. 1, 6, 10, in *Wikipedia Encyclopedia,* s.v. "Charles Waterhouse (artist)."

Preface:

1. *AZ Quotes,* s.v. "George S. Patton,"

Acknowledgments:

1. "Matthew 5:37," *Common English Bible,* Bible Gateway.
2. *Super-Squad: The Now Missing Component,* by H. John Poole (Emerald Isle, NC: Posterity Press, January 2020), parts two and three.

Chapter 1: *Accomplishing the Impossible*

1. "Return to Iwo Jima: 75 years After US-Japan Battle, What Is the Island Like Today," by Julian Ryall, *South China Morning Post,* 2 June 2019.
2. "Iwo Jima: The Most Horrific of All World War II Battles," from Warfare Hist. Network, *The National Interest,* 15 December 2018.
3. *The Third Marine Division,* by Robert A. Aurthur and Kenneth Cohlmia, ed. Robert T. Vance (Washington, D.C.: Infantry Journal Press, 1948), p. 244.
4. TF 56 Intel. Rpt. 3, in Lt.Col. Whitman S. Bartley, *Iwo Jima: Amphibious Epic* (Washington, D.C.: Hist. Branch, HQMC, 1954), p. 28.
5. 1st Supplement to Nanpo Shoto Information Bulletin No. 122-44, 10Oct44, CinCPac-CinCPOA Bulletin No. 9-45, 10Jan45, in Lt.Col. Whitman S. Bartley, *Iwo Jima: Amphibious Epic* (Washington, D.C.: Hist. Branch, HQMC, 1954), p. 28.

6. "Memories of Iwo Jima," interview of four Iwo Jima veterans, *Marines TV,* Youtube video, 26 March 2018.

7. "Western Pacific Operations," *History of Marine Corps Operations in World War II,* by George W. Garand and Truman R. Strowbridge, vol. IV, part VI, Hist. Branch, G-3 Div., HQMC, 1971, chapt. 4, p. 504.

8. Lt.Col. Whitman S. Bartley, *Iwo Jima: Amphibious Epic* (Washington, D.C.: Hist. Branch, HQMC, 1954), p. 203; *The Spearhead: The World War II History of the 5th Marine Division,* by Howard M. Connor (Washington, D.C.: Infantry Journal Press, 1950), p. 37.

9. "Iwo Jima and Okinawa: Death at Japan's Doorstep," *National WWII Museum* website, 10 July 2017.

10. "Return to Iwo Jima," by Ryall.

11. *The Third Marine Division,* by Aurthur and Cohlmia, p. 224.

12. *The Amphibians Came to Conquer: The Story of Admiral Richmond Kelly Turner,* by George Carroll Dyer, Dept. of the Navy, USMC, FMFRP 12-109-II, vol. 2, p. 1023; "The Black Sands of Iwo Jima," American Veteran's Ctr., n.d.; *Ennis,* by Laurie J. Wilson and Peggy Holland Rankin (Arcadia Publishing, February 2009), p. 111.

13. "Western Pacific Operations," by Garand and Strowbridge, vol. IV, part VI, chapt. 4, pp. 506-508; "Iwo Jima: The Most Horrific of All World War II Battles," from Warfare Hist. Network; Alexander, *Closing In,* p. 13.

14. Bartley, *Iwo Jima: Amphibious Epic,* p. 50; *The Spearhead* (5th Marine Division's unofficial website), s.v. "Battle of Iwo Jima"; Alexander, *Closing In,* p. 13.

15. "Western Pacific Operations," by Garand and Strowbridge, vol. IV, part VI, chapt. 4, p. 508.

15. Iwo Jima: The Most Horrific of All World War II Battles," from Warfare Hist. Network; "A Season In Hell: Iwo Jima 50th Anniversary," by Nathan Miller, *Baltimore Sun,* 19 February 1995.

17. Ibid.

18. "Western Pacific Operations," by Garand and Strowbridge, vol. IV, part VI, chapt. 4, p. 516.

19. Ibid.; "Iwo Jima: The Most Horrific of All World War II Battles," from Warfare Hist. Network.

20. "Western Pacific Operations," by Garand and Strowbridge, vol. IV, part VI, chapt. 1, p. 459.

21. *Wikipedia Encyclopedia,* s.v. "Planning for the Battle of Iwo Jima."

22. Ibid.

23. *The Spearhead: The World War II History of the 5th Marine Division,* by Howard M. Connor (Washington, D.C.: Infantry Journal Press, 1950), p. 38.

24. "Remember the Sacrifice of Iwo Jima Veterans," *Marines TV,* Youtube video, 14 March 2012.

25. "Iwo Jima: The Most Horrific of All World War II Battles," from Warfare Hist. Network.

26. Ibid; *Iwo Jima: The Account of the Epic Battle That Turned the Tide of World War II,* by Richard F. Newcomb (N.p: Griffin, May 2002), p. 15.

27. "Remember the Sacrifice of Iwo Jima Veterans," *Marines TV.*

28. "Western Pacific Operations," by Garand and Strowbridge, vol. IV, part VI, chapt. 1, p. 455.

29. *The Speadhead: WWII History of the 5th Marine Division,* by Connor, p. 35.

30. Ibid.

31. "To the Shores of Iwo Jima," U.S. Dept. of the Navy, Youtube video, 22 February 2017, with U.S. National Archives footage.

32. Navaho codetalker, in "Japanese Codetalkers," *In Search of History,* Hist. Channel, 30 March 1999.

33. *Spearhead: The WWII History of the 5th Marine Division,* by Connor, p. 35.

34. *Iwo Jima: The Account of the Epic Battle That Turned the Tide of World War II,* by Richard F. Newcomb (N.p: Griffin, May 2002), pp. 3-15; Alexander, *Closing In,* p. 45; "31st [Naval Construction Battalion] Spearhead" (n.p., n.d.); Bartley, *Iwo Jima: Amphibious Epic,* p. 125; "Geology and Ground-Water Resources of Iwo Jima," by Frank A. Swenson, GSA Bulletin (1948) 59 (10): 995–1008, 1 October 1948.

35. Memorandum for the record by H.J. Poole, after visiting an ancient hilltop desert fortress with rainwater collection system in Rajasthan around 1997.

36. Bartley, *Iwo Jima: Amphibious Epic,* p. 4; *Iwo Jima,* by Newcomb, pp. 3-15.

37. "Iwo Jima: The Unnecessary Battle," by Maj. Jonathan T. Baker, Master's Thesis (Quantico, VA: Cmd, and Staff College, Marine Corps Univ., 25 March 2011); *Investigating Iwo: The Flag Raisings in Myth, Memory, & Esprit de Corps,* ed. Breanne Robertson (Quantico, VA: Marine Corps Hist. Div., 2019), p. 35; *Iwo Jima,* by Newcomb, p. 15.

38. "Report of Intelligence: Iwo Jima Operation," Hdqts., Expeditionary Troops, Task Force 56, ACofS G-2, FMFPac memo to the CG of 1 April 1945.

39. *Weatherbase.com,* s.v. "Iwo Jima, Japan."

40. Ibid.

41. *The Spearhead: The WWII History of the 5th Marine Division,* by Connor, pp. 35-37.

42. *The Third Marine Division,* by Aurthur and Cohlmia, p. 238.

43. *So Sad to Fall in Battle: An Account of War Based on General Tadamichi,* by De Kumiko Kakehashi (Novato, CA: Presidio Press, 2009), reprint edition, map.

44. *Iwo Jima,* by Newcomb, p. 323.

45. Bartley, *Iwo Jima: Amphibious Epic,* chapt. 4.

46. *The Third Marine Division,* by Aurthur and Cohlmia, p. 224.

47. Bartley, *Iwo Jima: Amphibious Epic,* p. 6.

48. "Report of Intelligence: Iwo Jima Operation," by ACofS G-2, FMFPac, pp. 18-30; Bartley, *Iwo Jima: Amphibious Epic,* appendix VI.

49. Alexander, *Closing In*, p. 1.

50. Bartley, *Iwo Jima: Amphibious Epic,* appendix III; "Remember the Sacrifice of Iwo Jima Veterans," *Marines TV*; "Western Pacific Operations," by Garand and Strowbridge, vol. IV, part VI, appendix H; "Report of Intelligence: Iwo Jima Operation," by ACofS G-2, FMFPac, p. 30; Alexander, *Closing In*, p. 47.

Chapter 2: *Extent of Japanese Excavation*

1. *Super-Squad,* by Poole, chapts. 3, 8, and 13.

2. Philip Warner, *Japanese Army of World War II,* Men-at-Arms Series, color plates by Michael Youens (London: Osprey Publications Ltd., 1972), pp. 23-25.

3. "Western Pacific Operations," by Garand and Strowbridge, vol. IV, part VI, chapt. 11, pp. 692, 698.

4. Warner, *Japanese Army of World War II,* pp. 23-25; "Vehicle path and combat-shattered vegetation in the rugged terrain of northern Iwo Jima," WWII Database photo, 21 April 1945.

5. "Iwo Jima Journal: A Pacific Isle That Can't Quite Rest in Peace," by Sheryl Wudunn. *New York Times,* 15 March 1995.

6. Bill D. Ross, *Iwo Jima: Legacy of Valor* (New York: Vintage, 1986), p. 85; Alexander, *Closing In*, pp. 20, 28.

7. Ibid., pp. 20-24; Bartley, *Iwo Jima: Amphibious Epic*, p. 48.

8. Ross, *Iwo Jima: Legacy of Valor,* p. 249.

9. Ibid.; Bartley, *Iwo Jima: Amphibious Epic,* pp. 134-136.

10. Bartley, *Iwo Jima: Amphibious Epic*, pp. 149, 164.

11. Alexander, *Closing In,* p. 46.

12. Ross, *Iwo Jima: Legacy of Valor,* p. 217.

13. Ibid.; Bartley, *Iwo Jima: Amphibious Epic,* p. 149.

14. Bartley, *Iwo Jima: Amphibious Epic,* pp. 15, 159.

15. Ibid., p. 186.

16. Ibid., p. 189.

17. Ibid.. p. 15.

18. Ibid., pp. 14, 15, 146.

19. Ibid., pp. 140, 141.

20. *What Cameramen Saw Through Their Lenses,* 85:00, A Tribute to WWII Combat Cameramen of Japan Series (Tokyo: Nippon TV, 3 August 1995).

21. Alexander, *Closing In,* "The Drive North."

22. "To the Shores of Iwo Jima," U.S. Dept. of the Navy, 22 February 2017, with U.S. National Archives footage from Dept. of the Navy movie of that name in 1945, original unedited film may be at https://www.youtube.com/watch?v=4er8u9vd-Pw.

23. *What Cameramen Saw Through Their Lenses,* Nippon TV.

24. Bartley, *Iwo Jima: Amphibious Epic*, p. 15.

25. Keith Wheeler and the editors of Time-Life Books, *The Road to Tokyo,* World War II Series (Alexandria, VA: Time-Life Books, 1979), p. 42; "Sulfur Island," by P.J. O'Rourke, *The Atlantic,* June 2004; "Japanese Defenses and Fortifications Tawara, Iwo, Jima, and Okinawa 1943-1945," by J. David Rogers and Karl F. Hasselmann (Univ. of Missouri at Rolla, n.d.), p. 54.

26. "Operation Detachment: The Battle for Iwo Jima February - March 1945," *History of War,* n.d., "The Japanese Prepare," as retrieved in March 2020 from the following url: http://www.historyofwar.org/articles/battles_iwojima.html#d3.

27. Alexander, *Closing In,* p. 5.

28. Bartley, *Iwo Jima: Amphibious Epic*, p. 184.

29. Ibid., pp. 149, 150.

30. *Wikipedia Encyclopedia,* s.v. "Planning for the Battle of Iwo Jima"; "The WW II Japanese Army on Iwo Jima had 1000 Korean laborers building the caves, tunnels and fortifications," by Grant Lee, *Quora,* 2 June 2017.

31. Alexander, *Closing In,* pp. 46, 47.

32. Ibid., p. 46.

33. Ross, *Iwo Jima: Legacy of Valor*, p. 331.

34. Alexander, *Closing In,* pp. 46, 47.

35. Bartley, *Iwo Jima: Amphibious Epic,* p. 189.

36. "The 147th Infantry Regiment at Iwo Jima," by Douglas E. Nash, Sr., *Army History Magazine,* Fall 2017, U.S. Army Ctr. of Mil. Hist., pp. 7-19.

37. *So Sad to Fall in Battle,* by Kakehashi, p. 64.

38. Alexander, *Closing In,* pp. 46, 47.

39. "The 147th Infantry Regiment at Iwo Jima," by Nash.

40. "Japanese Surrender After Four Year Hiding," *Pacific Stars and Stripes*. 10 January 1949. p. 5; Donald Cook, "Capture of Two Holdouts January 6, 1949," *No Surrender: Japanese Holdouts*, n.d.; in *Wikipedia Encyclopedia,* s.v. "Battle of Iwo Jima."

41. "Report of Intelligence: Iwo Jima Operation," by ACofS G-2, FMFPac, map.

42. "Japanese Defenses and Fortifications Tawara, Iwo, Jima, and Okinawa 1943-1945," by J. David Rogers and Karl F. Hasselmann (Univ. of Missouri at Rolla, n.d.), p. 53.

43. "Situation Map of Iwo Jima," as retrieved in February 2020 from the following url: https://bostonraremaps.com/inventory/situation-map-iwo-jima-1944/.

44. "Report of Intelligence: Iwo Jima Operation," by ACofS G-2, FMFPac, p. 37.

45. Alexander, *Closing In,* p. 45.

Chapter 3: *A State-of-the-Art Japanese Defense*

1. President John F. Kennedy, quote from his 1962 West Point graduation speech, in "Unconventional Warfare Fundamentals," at *The Irregular Warrior* website.

2. *The Tigers's Way: A U.S. Private's Best Chance at Survival,* by H. John Poole (Emerald Isle, NC: Posterity Press, 2003), chapt. 12.

3. Alexander, *Closing In,* p. 13.

4. *Wikipedia Encyclopedia,* s.v. "John Basilone (Navy Cross citation)."

5. Ibid.

6. *I'm Staying with My Boys: The Heroic Life of Sgt. John Basilone, USMC,* by Jim Prosser with Jerry Cutter [Basilone nephew] (New York: St. Martin's Griffin, 2004), pp. 91-94.

7. Merritt E. Benson, "Seventeen Days at Iwo Jima," *Marine Corps Gazette,* February 2000, p. 32.

8. Alexander, *Closing In,* p. 12.

9. Ibid., p. 14.

10. Ibid., p. 19.

11. Ibid., pp. 23, 24.

12. Ibid., p. 23.

13. Ibid.

14. *The Spearhead* (5th Marine Division's unofficial website), s.v. "The Battle of Iwo Jima,"

15. *Super-Squad,* by Poole, chapt. 14.

16. Alexander, *Closing In,* p. 23.

17. Ibid.

18. Ross, *Iwo Jima: Legacy of Valor*, pp. 72, 73.

19. *The Spearhead: The WWII History of the 5th Marine Division,* by Connor, pp. 48, 49; Alexander, *Closing In,* pp. 17, 23.

20. *Gung Ho: The Corps' Most Progressive Tradition,* by H. John Poole (Emerald Isle, NC: Posterity Press, 2012), chapt. 10.

21. "Into the Meatgrinder," chapt 6, of *Battle for Iwo Jima,* by Robert Leckie (N.p.: I Books, 2004).

22. *Super-Squad,* by Poole, p. 21.
23. Bartley, *Iwo Jima: Amphibious Epic,* pp. 15, 96, 149.
24. Alexander, *Closing In,* p. 28.
25. "Battle of Iwo Jima 19 Feb 1945 - 26 Mar 1945," by C. Peter Chen, *World War II Database,* n.d.; *I'm Staying with My Boys,* by Jim Prosser. epilogue.
26. "Reflections on Iwo Jima: The Japanese Tunnels," interview of Col. Dave Severance USMC (Ret.), CO of E/2/25, *Marines TV,* Youtube video, 26 March 2011.
27. Bartley, *Iwo Jima: Amphibious Epic,* p. 154.
28. "Into the Meatgrinder," chapt. 6, *Battle for Iwo Jima,* by Leckie.
29. *Super-Squad,* by Poole, chapt. 8.
30. Alexander, *Closing In,* p. 23.
31. "Into the Meatgrinder," chapt. 6, *Battle for Iwo Jima,* by Leckie.
32. Ibid.
33. *Wikipedia Encyclopedia,* s.v. "Tony Stein"; Ross, *Iwo Jima: Legacy of Valor,* pp. 70, 71.
34. Alexander, *Closing In,* p. 40.

Chapter 4: *How Far Had Corps Small-Unit Maneuvers Evolved*

1. *Super-Squad,* by Poole, chapt. 6.
2. Bartley, *Iwo Jima: Amphibious Epic,* p. 141.
3. *The Third Marine Division,* by Aurthur and Cohlmia, p. 226.
4. "Jack Lucas and the Battle of Iwo Jima." *The History Guy,* Youtube video presentation, 26 February 2020; *I'm Staying with My Boys,* by Prosser, pp. 93, 316, 317.
5. *I'm Staying with My Boys,* by Prosser, pp. 91-94; *Sands of Iwo Jima* (Hollywood: Republic Pictures, 1988), videocassette.
6. *Home of the Heroes,* s.v. "Max E. Melville."
7. Ibid., s.v. "Ross F. Gray."
8. Ibid., s.v. "Martin J. Queeney."
9. "The Battle of Iwo Jima—Marines Remember Their Heritage," interview of Col. Dave Severance USMC (Ret.), CO of E/2/25. *Marines TV,* YouTube video, 4 March 2011.
10. "Into the Meatgrinder," chapt. 6, of *Battle for Iwo Jima,* by Leckie.
11. Alexander, *Closing In,* p. 27.
12. *Home of the Heroes,* s.v. "John K. Wells."
13. "Marine Who Led Famous Charge up Mountain on Iwo Jima Remembered," *USMC Life,* 18 February 2016; "Excellence in Land Navigation: The 1stLt. Keith Wells Award," Marine Corps Reserve Assoc. website.
14. *Wikipedia Encyclopedia,* s.v. "Joseph J. McCarthy."

15. *Home of the Heroes,* s.v. "Paul B. Balducci."

16. "Marine Who Led Famous Charge up Mountain on Iwo Jima Remembered"; *Give Me Fifty Marines Not Afraid to Die,* by John K. Wells (N.p.: Ka-Well Enterprises, 1995), editorial review at amazon.com, "about the author."

17. Bartley, *Iwo Jima: Amphibious Epic,* pp. 140, 141.

18. Ibid., p. 160.

19. Alexander, *Closing In,* p. 27.

20. Bartley, *Iwo Jima: Amphibious Epic,* p. 154.

21. Ibid., pp. 107, 110, 147.

22. Ibid., p. 171.

23. Company C, 1st Battalion, 23rd Marines website, "Approach and Capture of Iwo Jima."

24. *Home of the Heroes,* s.v. "Joe R. Crow."

25. Ibid., s.v. "Edward J. Sivicky."

26. *The Spearhead: The WWII History of the 5th Marine Division,* by Connor, pp. 48, 49.

27. Bill Rose, as quoted in "A Season In Hell," by Miller.

28. Bartley, *Iwo Jima: Amphibious Epic,* pp. 128, 129; "Into the Meatgrinder," chapt. 6, of *Battle for Iwo Jima,* by Leckie.

29. *The First Battalion of the 28th Marines on Iwo Jima: A Day-by-Day History from Personal Accounts and Official Reports, with Complete Muster Rolls,* by Robert E. Allen (London: McFarland & Company, 1999), pp. 91, 92.

30. Maj.Gen. Edward J. Miller USMC [in replacement draft at Iwo Jima as a new lieutenant], in conversation with author around 1975.

31. *Gung Ho,* by Poole, chapt. 9.

32. *I'm Staying with My Boys*, by Prosser with Cutter [Basilone, nephew], pp. 91-94.

33. *Wikipedia Encyclopedia*, s.v. "Harry B. Liversedge."

34. *Investigating Iwo: The Flag Raisings in Myth, Memory, & Esprit de Corps,* ed. Breanne Robertson (Quantico, VA: Marine Corps Hist. Div., 2019), pp. 153, 154; *The United States Marines on Iwo Jima: The Battle and the Flag Raisings,* by Bernard C. Nalty and Danny J. Crawford (Washington, D.C.: Hist. and Museums Div., HQMC, 1995), *Wikipedia Encyclopedia*, s.v. "Michael Strank."

35. *The First Battalion of the 28th Marines on Iwo Jima,* by Allen, pp. 91, 92.

36. *World War II Medal of Honor Recipients* (London: Osprey, n.d.), p. 51.

37. Bartley, *Iwo Jima: Amphibious Epic,* p. 93.

38. Ibid., pp. 92, 171.

39. "The 147th Infantry Regiment at Iwo Jima," by Nash, p. 15.

40. "Jack Lucas and the Battle of Iwo Jima," Youtube video presentation.

41. *The Spearhead: The WWII History of the 5th Marine Division,* by Connor, p. 51.
42. *Home of the Heroes,* s.v. "Merritt M. Savage."
43. *Wikipedia Encyclopedia*, s.v. "Battle of Bougainville" and "3rd Marine Division."
44. Bartley, *Iwo Jima: Amphibious Epic,* p. 111
45. *The Spearhead: The WWII History of the 5th Marine Division,* by Connor, p. 23.
46. *Wikipedia Encyclopedia*, s.v. "Hershel W. Williams."
47. Alexander, *Closing In,* p. 43.
48. Ibid., pp. 43-45.

Chapter 5: *Degree of Offensive Japanese Activity on Iwo*

1. John Shaw and the editors of Time-Life Books, *Red Army Resurgent* (Chicago: Time-Life Books, 1979), p. 167.
2. "Iwo Jima: The Most Horrific of All World War II Battles," from Warfare Hist. Network.
3. Bartley, *Iwo Jima: Amphibious Epic,* p. 29.
4. *Guardian Joe: How Less Force Helps the Warrior,* by H. John Poole (Emerald Isle, NC: Posterity Press, 2018), chapt. 9.
5. Report of Intelligence: Iwo Jima Operation," ACofS G-2, FMFPac, G-2 Periodic Rpt. #2.
6. Alexander, *Closing In,* p. 20.
7. "Western Pacific Operations," by Garand and Strowbridge, vol. IV, part VI, chapt. 4, pp. 526, 527.
8. Bartley, *Iwo Jima: Amphibious Epic,* pp. 59, 65, 66.
9. Ibid., p. 83, 84
10. Report of Intelligence: Iwo Jima Operation," ACofS G-2, FMFPac, G-2 Periodic Rpt. #3.
11. *The Spearhead* (5th Marine Division's unofficial website), s.v. "The Battle of Iwo Jima" D-Day to D+10.
12. Bartley, *Iwo Jima: Amphibious Epic,* p. 87.
13. Report of Intelligence: Iwo Jima Operation," ACofS G-2, FMFPac, G-2 Periodic Rpt. #4;
14. Bartley, *Iwo Jima: Amphibious Epic,* p. 76.
15. "Western Pacific Operations," by Garand and Strowbridge, vol. IV, part VI, chapt. 5, p. 538.
16. *The Spearhead* (5th Marine Division's unofficial website), s.v. "The Battle of Iwo Jima" D-Day to D+10.
17. Bartley, *Iwo Jima: Amphibious Epic,* p. 87; *Guardian Joe,* by Poole, chapts. 19, 23.
18. Report of Intelligence: Iwo Jima Operation," ACofS G-2, FMFPac, G-2 Periodic Rpt. #5.

19. Ibid., G-2 Periodic Rpt. #6.

20. Bartley, *Iwo Jima: Amphibious Epic,* p. 78.

21. "Western Pacific Operations," by Garand and Strowbridge, vol. IV, part VI, chapt. 5, p. 539.

22. *The Spearhead* (5th Marine Division's unofficial website), s.v. "The Battle of Iwo Jima" D-Day to D+10.

23. Bartley, *Iwo Jima: Amphibious Epic,* p. 91.

24. Ibid., p. 93.

25. Ibid., p. 12.

26. *Guardian Joe,* by Poole, p. 230.

27. *Investigating Iwo,* ed. Robertson, pp. 24-26.

28. Bartley, *Iwo Jima: Amphibious Epic,* p. 91, map XX.

29. Alexander, *Closing In,* p. 1.

30. *Home of the Heroes,* s.v. "Andrew J. Carter."

31. *The Third Marine Division,* by Aurthur and Cohlmia, p. 238.

32. Alexander, *Closing In,* p. 28.

33. Bartley, *Iwo Jima: Amphibious Epic*, pp. 172-174.

34. *The Third Marine Division,* by Aurthur and Cohlmia, p. 238.

35. "Western Pacific Operations," by Garand and Strowbridge, vol. IV, part VI, chapt. 11, pp. 685-697.

36. Ibid., p. 699.

37. Bartley, *Iwo Jima: Amphibious Epic,* p. 176.

38. 4thMarDiv D-2 Periodic Rpt. No. 63, 6 Mar 45, in Bartley, *Iwo Jima: Amphibious Epic,* pp. 176, 177.

39. *Investigating Iwo,* ed. Robertson, appendix G, pp. 332-333.

40. Ibid., chapt. 1, p. 22.

41. Report of Intelligence: Iwo Jima Operation," ACofS G-2, FMFPac, p. 14.

42. *Super-Squad,* by Poole, chapt. 9.

43. *Handbook on Japanese Military Forces,* TM 30-480 (Washington, D.C.: U.S. War Dept., 1944), p. 117.

Chapter 6: *The Japanese Soldier's Code of Conduct*

1. *The Oxford Companion to World War II,* ed. I.C.B. Dear and M.R.D Foot (Oxford, U.K,: Oxford Univ. Press, 1995), p. 476, in *Wikipedia Encyclopedia*. s.v. "*Senjinkun* military code."

2. "Gyokusai or 'Shattering like a Jewel': Reflection on the Pacific War," by Hiroaki Sato, *Asia-Pacific Journal,* 1 February 2008.

3. John Toland, *The Rising Sun: The Decline and Fall of the Japanese Empire 1936-1945* (New York: Random House, 1970), p. 512, in *Wikipedia Encyclopedia*. s.v. "*Senjinkun* military code."

4. *Wikipedia Encyclopedia*. s.v. "*Senjinkun* military code."

5. John W. Dower, *War Without Mercy: Race and Power in the Pacific War*, Pantheon, 1987, in *Wikipedia Encyclopedia*. s.v. *"Senjinkun* military code."

6. John W. Dower, *War Without Mercy: Race and Power in the Pacific War* (New York, Pantheon, 1987), p. 26.

7. *Wikipedia Encyclopedia*, s.v. "Imperial Rescript to Soldiers and Sailors."

8. Ibid.

9. Stephen Turnbull, *Samurai: The World of the Warrior* (New York: Osprey Publishing, 2003), in *Wikipedia Encyclopedia*. s.v. *"Bushido."*

10. "Russo-Japanese War: Japan's First Big Surprise," by Robert M. Citino, *History Net,* n.d.; Oleg Benesch, *Inventing the Way of the Samurai: Nationalism, Internationalism, and Bushido in Modern Japan* (Oxford: Oxford Univ, Press, 2014), in *Wikipedia Encyclopedia*, s.v. *"Bushido."*

11. *Handbook on Japanese Military Forces*, p. 91.

12. Ibid., p. 98.

13. Ibid., p. 86.

14. Ibid., p. 85.

Chapter 7: *Japanese Homeland Assistance for Iwo Jima*

1. "The Sinking of the Japanese Submarine I-1 off of Guadalcanal and the Recovery of Its Secret Documents," by Netisha, posted in *Military Records, Navy, Marines, & Coast Guard, World War II,* by Dr. Greg Bradsher, Senior Archivist at the National Archives at College Park on 31 October 2017.

2. *Wikipedia Encyclopedia,* s.v. "Operation Ke."

3. Ibid., s.v. "Japanese occupation of Kiska."

4. Ibid., s.v. "Operation Cottage."

5. Bartley, *Iwo Jima: Amphibious Epic,* p. 27.

6. Ibid., pp. 186. 187.

7. Ibid., p. 114.

8. "IJN Submarine I-44: Tabular Record of Movement," by Bob Hackett & Sander Kingsepp, n.d., as retrieved in March 2020 from the following url: http://www.combinedfleet.com/I-44.htm.

9. Ibid.

10. Ibid.

11. Jerry Gray (retired U.S. Navy Chief Petty Officer), in conversation with author around January 2020.

12. *Wikipedia Encyclopedia,* s.v. "I-400 Class Submarine"; *Sunk,* by Mochitsura Hashimoto, (N.p.: Henry Holt and Co., 1954), p. 213, in op.cit.

13. "Iwo Jima: The Most Horrific of All World War II Battles," from Warfare Hist. Network.

14. *Wikipedia Encyclopedia,* s.v. "Submarines of the Imperial Japanese Navy."

15. "Western Pacific Operations," by Garand and Strowbridge, vol. IV, part VI, chapt. 1, p. 456; *Spearhead: The WWII History of the 5th Marine Division,* by Connor, p. 35.

16. *So Sad to Fall in Battle,* by Kakehashi,

17. "The 147th Infantry Regiment at Iwo Jima," by Nash.

18. Bartley, *Iwo Jima: Amphibious Epic,* p. 183.

19. "The 147th Infantry Regiment at Iwo Jima," by Nash.

20. "Return to Iwo Jima," by Ryall.

21. *Iwo Jima,* by Newcomb, p. 13.

22. Bartley, *Iwo Jima: Amphibious Epic,* pp. 191, 192.

23. Ibid.; "The 147th Infantry Regiment at Iwo Jima," by Nash.

Chapter 8: *Procedural Challenges for the Marines on Iwo*

1. *Wikipedia Encyclopedia,* s.v. "John Basilone."

2. Ibid., s.v. "Battle of Monte Cassino."

3. "Western Pacific Operations," by Garand and Strowbridge, vol. IV, part VI, chapt. 11, p. 695.

4. Bartley, *Iwo Jima: Amphibious Epic,* p. 206.

5. Alexander, *Closing In*, p. 28.

6. Bartley, *Iwo Jima: Amphibious Epic,* p. 167.

7. *The Third Marine Division,* by Aurthur and Cohlmia, p. 226.

8. Ross, *Iwo Jima: Legacy of Valor,* p. 163.

9. Bartley, *Iwo Jima: Amphibious Epic,* p. 149; *Handbook on Japanese Military Forces,* TM 30-480, p. 34; "Japanese Organisation in WWII," as retrieved in November 2018 from this url: http://www-solar.mcs.st-and.ac.uk/~aaron/WW2ORG/jap.html.

10. *Wikipedia Encyclopedia,* s.v. "John Basilone."

11. Ibid., s.v. "Tony Stein"; *Investigating Iwo,* ed. Robertson, pp. 17, 18.

12. Alexander, *Closing In*, p. 12.

13. Ibid., pp. 22, 23.

14. VACRpt, AnnexC, Appendix10, p. 33, in Bartley, *Iwo Jima: Amphibious Epic,* p. 196.

15. Bartley, *Iwo Jima: Amphibious Epic,* p. 154.

16. *Wikipedia Encyclopedia,* s.v. "Tony Stein," MoH citation.

17. Alexander, *Closing In*, p. 23.

18. Bartley, *Iwo Jima: Amphibious Epic,* p. 36

19. Ibid., appendix III; *The Spearhead: The World War II History of the 5th Marine Division,* by Connor, p. 53.

20. Company C, 1st Battalion, 23rd Marines website, "Approach and Capture of Iwo Jima."

21. Ibid.

22. "The Legend of Suicide Charley: Company C, 1st Battalion, 7th Marines, and the Battle for Henderson Field, 24–25 October 1942," by Major Gary Cozzens, USMCR (Ret.), *Marine Corps History,* Summer 2017, p. 39.

23. *I'm Staying with My Boys*, by Prosser and Cutter [Basilone nephew], pp. 178, 179.

24. Ibid., pp. 35-41.

25. Ibid., pp. 297, 298.

26. Ibid., pp. 191-194.

27. Ibid., p. 327.

Chapter 9: *Up to the Flag-Raising on Suribachi*

1. *Wikipedia Encyclopedia,* s.v. "John Basilone."

2. *Home of the Heroes,* s.v. "Joshua S. Bell."

3. "Report of Intelligence: Iwo Jima Operation," by ACofS G-2, FMFPac, p. 36.

4. Ibid., appendix A.

5. Ibid., p. 37.

6. Ross, *Iwo Jima: Legacy of Valor*, pp. 72, 73; Bartley, *Iwo Jima: Amphibious Epic,* p. 56.

7. *Home of the Heroes,* s.v. "Darrell Samuel Cole."

8. Alexander, *Closing In*, p. 23.

9. *Home of the Heroes*, s.v. "Camille J. Doiron."

10. Ibid., s.v. "Edward J. Sivicky."

11. Ibid., s.v. "James K. Vincent."

12. "Report of Intelligence: Iwo Jima Operation," ACofS G-2, FMFPac, p. 35; *Investigating Iwo,* ed. Robertson, p. 17.

13. *Home of the Heroes,* s.v. "Conrad F. Shaker."

14. *Wikipedia Encyclopedia,* s.v. "Tony Stein."

15. Ross, *Iwo Jima: Legacy of Valor,* pp. 70, 71.

16. *Home of the Heroes,* s.v. "William A. Goff."

17. Alexander, *Closing In*, p. 20.

18. *Home of the Heroes,* s.v. "Raymond W. Bahring."

19. *Wikipedia Encyclopedia,* s.v. "Robert Hugo Dunlap."

20. *Home of the Heroes,* s.v. "John B. Babich."

21. "Marine Who Led Famous Charge up Mountain on Iwo Jima Remembered," *Marine Corps Times,* by Matthew Schehl, 17 February 2016.

22. *Home of the Heroes,* s.v. "Harold G. Schrier."

23. Alexander, *Closing In*, p. 27.

24. "Excellence in Land Navigation: The 1stLt. Keith Wells Award," Marine Corps Reserve Assoc. website.

25. *Sands of Iwo Jima,* Republic Pictures.

26. *Wikipedia Encyclopedia,* s.v. "Michael Strank."

Chapter 10: *Through the Second Defense Belt*

1. *Warfighting,* MCDP 1 (Washington, D.C.: HQMC, 1997), pp. 42, 43.

2. Alexander, *Closing In,* p. 40.

3. "Report of Intelligence: Iwo Jima Operation," ACofS G-2, FMFPac, p. 35.

4. Ibid., p. 33.

5. Ibid., p. 16; Bartley, *Iwo Jima: Amphibious Epic,* p. 88.

6. Bartley, *Iwo Jima: Amphibious Epic,* any map and pp. 33, 34.

7. *The Third Marine Division,* by Aurthur and Cohlmia, p. 224.

8. Ibid., p. 232.

9. *The Spearhead: The WWII History of the 5th Marine Division,* by Connor, pp. 89, 91.

10. Bartley, *Iwo Jima: Amphibious Epic,* p. 96.

11. "Japanese Defenses and Fortifications Tawara, Iwo, Jima, and Okinawa 1943-1945," by Rogers and Hasserman.

12. *The Third Marine Division,* by Aurthur and Cohlmia, p. 231.

13. *Home of the Heroes,* s.v. "William L. Shannon."

14. *Marines in the Central Solomons,* by Maj. John N. Rentz, "USMC Historical Monograph" (Washington, D.C.: Hist. Br., HQMC, 1952), chapt. 4, "From Rice to Bairoko, pp. 113, 114; Bartley, *Iwo Jima: Amphibious Epic,* p. 85.

15. Bartley, *Iwo Jima: Amphibious Epic,* p. 158.

16. Ibid., p. 107.

17. Ibid., p. 11.

18. Ibid., p. 15.

19. Western Pacific Operations," by Garand and Strowbridge, vol. IV, part VI, chapt. 9, pp. 635-637; *Iwo Jima,* by Newcomb, "Place Names on Iwo Jima" appendix, p. 323; *Investigating Iwo,* ed. Robertson, p. 18.

20. *Home of the Heroes,* s.v. "John H. Wherry."

21. *Wikipedia, Encyclopedia,* s.v. " William G. Walsh"

22. *Home of the Heroes,* s.v. "Harry Towne."

23. Bartley, *Iwo Jima: Amphibious Epic,* pp. 128, 129.

24. *Home of the Heroes,* s.v. "Donn J. Robertson."

25. Ibid., s.v. "Philip R. Gray."

26. *Strategic Rifleman: Key to More Moral Warfare,* by H. John Poole (Emerald Isle, NC: Posterity Press, 2014), chapt. 17; *The Spearhead* (5th Marine Division's unofficial website), s.v. "The Battle of Iwo Jima."

27. *Home of the Heroes,* s.v. "Raymond W. Bahring."

28. *So Sad to Fall in Battle,* by Kakehashi, map; "Report of Intelligence: Iwo Jima Operation," by ACofS G-2, FMFPac, map.

29. *The Third Marine Division,* by Aurthur and Cohlmia, p. 231.

30. *Home of the Heroes,* s.v. "Raoul J. Archambault."

31. "Western Pacific Operations," by Garand and Strowbridge, vol. IV, part VI, chapt. 7, pp. 574-577; *The Third Marine Division,* by Aurthur and Cohlmia, pp. 231, 232.

32. *Home of the Heroes,* s.v. "Carl J. Linblad."

33. Ibid., s.v. "Albert M. Villa."

34. Ibid., s.v. "Orville O. Vaught."

35. Ibid., s.v. "Robert Fransko."

36. *Wikipedia, Encyclopedia,* s.v. "Wilson T. Watson."

37. *The Third Marine Division,* by Aurthur and Cohlmia, p. 232.

38. *Home of the Heroes,* s.v. " Joseph J. McCarthy."

39. Ibid., s.v. "Dominick J. Grossi."

40. Ibid., s.v. "Harold C. Boehm."

41. Company C, 1st Battalion, 23rd Marines website, "Approach and Capture of Iwo Jima."

42. Bartley, *Iwo Jima: Amphibious Epic,* p. 149.

43. "Into the Meatgrinder," chapt. 6, of *Battle for Iwo Jima,* by Leckie.

44. *Wikipedia, Encyclopedia,* s.v. "Douglas T. Jacobson."

45. "Assessing the Conventional Balance: The 3:1 Rule and Its Critics," by John J. Mearsheimer, *International Security,* vol. 13, no. 4, Spring 1989, pp. 54-89.

46. Bartley, *Iwo Jima: Amphibious Epic,* maps I-XXI; "Into the Meatgrinder," chapt. 6, of *Battle for Iwo Jima,* by Leckie.

47. *From Uraga to San Francisco: A Century of Japanese Diplomacy, 1853-1952,* by Hisahiko Okazaki (Japan: Echo, 2007), p. 212.

48. *Home of the Heroes,* s.v. "Joshua S. Bell."

49. Western Pacific Operations," by Garand and Strowbridge, vol. IV, part VI, chapt. 10, pp. 662-663; Company C, 1st Battalion, 23rd Marines website, "Approach and Capture of Iwo Jima."

50. Ibid.

51. *Home of the Heroes,* s.v. "Henry T. Pound."

52. "Into the Meatgrinder," chapt 6, of *Battle for Iwo Jima,* by Leckie; 4thMarDiv D-3 Periodic Rpt. No. 57, 2 March 45, in in Bartley, *Iwo Jima: Amphibious Epic,* p. 164.

53. Western Pacific Operations," by Garand and Strowbridge, vol. IV, part VI, chapt. 10, p. 665; Bartley, *Iwo Jima: Amphibious Epic,* p. 165.

54. *Home of the Heroes*, s.v. "Frank J. Gambino."

55. "Into the Meatgrinder," chapt 6, of *Battle for Iwo Jima,* by Leckie.

56. Western Pacific Operations," by Garand and Strowbridge, vol. IV, part VI, chapt. 10, pp. 656-682; "Into the Meatgrinder," chapt 6, of *Battle for Iwo Jima,* by Leckie.

57. Ibid.

58. *Home of the Heroes,* s.v. "Orville W. Hahn."

59. Western Pacific Operations," by Garand and Strowbridge, vol. IV, part VI, chapt. 10, pp. 656-682.

60. Ibid.

61. *From Uraga to San Francisco,* by Okazaki, p. 212.

62. Alexander, *Closing In,* pp. 28, 35.

63. Ibid., p. 28.

64. Western Pacific Operations," by Garand and Strowbridge, vol. IV, part VI, chapt. 10, pp. 656-682.

65. Alexander, *Closing In,* pp. 28, 35.

66. 4thMarDiv D-2 Periodic Rpt. No. 68, 11Mar45, in Bartley, *Iwo Jima: Amphibious Epic,* p. 183.

67. *Home of the Heroes,* s.v. "John G. Ioanna."

68. 4thMarDiv D-2 Periodic Rpt. No. 69, 12Mar45, in Bartley, *Iwo Jima: Amphibious Epic,* p. 184.

69. Bartley, *Iwo Jima: Amphibious Epic,* p. 184.

70. *Wikipedia Encyclopedia,* s.v. "Francis Pierce, Jr."

71. Bartley, *Iwo Jima: Amphibious Epic.* p. 12.

72. Ibid., p. 184; Alexander, *Closing In,* p. 28.

73. "Western Pacific Operations," by Garand and Strowbridge, vol. IV, part VI, chapt. 5, p. 539.

74. Alexander, *Closing In,* pp. 1, 2.

75. Bartley, *Iwo Jima: Amphibious Epic,* p. 191.

76. "Sun Tzu Quotes," *Goodreads,* n.d.

Chapter 11: *With Regard to the Third Defense Belt and Gorge*

1. Bartley, *Iwo Jima: Amphibious Epic,* pp. 11, 187.

2. Alexander, *Closing In*, p. 31.

3. *So Sad to Fall in Battle,* by Kakehashi, map.

4. Bartley, *Iwo Jima: Amphibious Epic,* p. 11.

5. Western Pacific Operations," by Garand and Strowbridge, vol. IV, part VI, chapt. 10, p. 677.

6. *Handbook on Japanese Military Forces,* TM-E 30-480, p. 34; Bartley, *Iwo Jima: Amphibious Epic,* pp. 172 and 173.

7. *Home of the Heroes,* s.v. "Daris W. Latch."

8. Bartley, *Iwo Jima: Amphibious Epic*, pp. 14, 15, 146; "The 147th Infantry Regiment at Iwo Jima," by Nash, p. 16.

9. Bartley, *Iwo Jima: Amphibious Epic*, pp. 186 and 187.

10. *The Spearhead* (5th Marine Division's unofficial website), s.v. "The Battle of Iwo Jima"; Bartley, *Iwo Jima: Amphibious Epic*, p. 186.

11. "Operation Detachment: The Battle for Iwo Jima February - March 1945," *History of War,* n.d., "D+20 - D+35: 11 March 1945 - 26 March 1945," as retrieved in April 2020 from the following url: http://www.historyofwar.org/articles/battles_iwojima.html#d20; Alexander, *Closing In,* p. 46; Bartley, *Iwo Jima: Amphibious Epic,* appendix VI; *The United States Marines on Iwo Jima,* by Nalty and Crawford. p. 21.

12. Bartley, *Iwo Jima: Amphibious Epic*, p. 190.

13. *Home of the Heroes,* s.v. "Bernhardt L. Mueller."

14. Ibid., s.v. "Michael R. Kost."

15. Ibid., s.v. "Calvin J. Bleau."

16. Ibid., s.v. "Oral L. Tankersley."

17. Ibid., s.v. Ray E. Wilson, Jr."

18. Ibid., s.v. "Henry A. Harrison."

19. Ibid., s.v. "Hubert J. Faltyn."

20. *Wikipedia Encyclopedia,* s.v. "Franklin Sigler."

21. *Home of the Heroes,* s.v. " Frank L. Palmer."

22. *The Spearhead* (5th Marine Division's unofficial website), s.v. "The Battle of Iwo Jima"

23. *Home of the Heroes*, s.v. "James T. Little."

24. Alexander, *Closing In,* p. 46.

25. *Home of the Heroes,* s.v. "Daniel C. Pollock."

26. Ibid., s.v. "Edward James Coleman."

27. *The Spearhead* (5th Marine Division's unofficial website), s.v. "The Battle of Iwo Jima," D+11 to D+20; Bartley, *Iwo Jima: Amphibious Epic*, maps X and XI.

28. *Home of the Heroes*, s.v. "Frank C. Caldwell."

29. Ibid., s.v. "M.L. Gelshenen."

30. *Wikipedia Encyclopedia*, s.v. "George Edward Gehlen."

31. Bartley, *Iwo Jima: Amphibious Epic,* pp. 108, 109.

32. "Western Pacific Operations," by Garand and Strowbridge, vol. IV, part VI, chapt. 7, pp. 583-590.

33. Ibid.; Alexander, *Closing In,* pp. 43-45; Bartley, *Iwo Jima: Amphibious Epic,* pp. 118, 119.

34. Lt.Col. Harold C. Boehm, in Bartley, *Iwo Jima: Amphibious Epic,* p. 119.

35. Western Pacific Operations," by Garand and Strowbridge, vol. IV, part VI, chapt. 7, pp. 590, 591.

36. Ibid.

37. Ibid., p. 592.

38. Bartley, *Iwo Jima: Amphibious Epic,* p. 178.

39. Ibid., p. 178.

40. Ibid., pp. 178-181.

41. Western Pacific Operations," by Garand and Strowbridge, vol. IV, part VI, chapt. 11, p. 688.

42. *The Spearhead* (5th Marine Division's unofficial website), s.v. "The Battle of Iwo Jima," D+11 to D+20.

43. "Western Pacific Operations," by Garand and Strowbridge, vol. IV, part VI, chapt. 4, p. 516; "Iwo Jima: The Most Horrific of All World War II Battles," from Warfare Hist. Network; "Remember the Sacrifice of Iwo Jima Veterans," *Marines TV;* "Report of Intelligence: Iwo Jima Operation," by ACofS G-2, FMFPac. p. 30; Alexander, *Closing In,* p. 2.

44. 3rd MarDiv AR, p. 17, in "Western Pacific Operations," by Garand and Strowbridge, vol. IV, part VI, chapt. 7, p. 592.

45. "Absolutely Brutal (And Smart) Tactics In Native American Warfare," by Justin Andress, *Ranker,* 4 February 2019.

46. Western Pacific Operations," by Garand and Strowbridge, vol. IV, part VI, chapt. 7, pp. 590, 591.

47. *The Spearhead* (5th Marine Division's unofficial website), s.v. "The Battle of Iwo Jima," D+11 to D+20.

Chapter 12: *During the Final Consolidation*

1. "Western Pacific Operations," by Garand and Strowbridge, vol. IV, part VI, chapt. 11, p. 711.

2. "Battle of Iwo Jima," by Joseph Alexander, *History Net,* n.d.; "Western Pacific Operations," by Garand and Strowbridge, vol. IV, part VI, chapt. 5, p. 539.

3. Alexander, *Closing In,* pp. 1, 2.

4. "The 147th Infantry Regiment at Iwo Jima," by Nash, p. 16.

5. Ibid.

6. "Western Pacific Operations," by Garand and Strowbridge, vol. IV, part VI, chapt. 11, p. 711.

7. "Into the Meatgrinder," chapt 6, of *Battle for Iwo Jima,* by Leckie.

8. Bartley, *Iwo Jima: Amphibious Epic,* p. 184.

9. Ibid. pp. 178-181.

10. Ibid,, p. 37.

11. "Report of Intelligence: Iwo Jima Operation," by ACofS G-2, FMFPac, p. 31.

12. "The 147th Infantry Regiment at Iwo Jima," by Nash, pp. 15, 16.

13. Ibid., pp. 16, 17.

14. "Japanese Surrender After Four Year Hiding," *Pacific Stars and Stripes,* 10 January 1949, p. 5; and "Profiles of Known Japanese Holdouts." by Yamakage Kufuku" (N.p.: Wanpela, 2007); both in *Wikipedia Encyclopedia,* s.v. "Japanese Holdout."

15. Facebook.com, s.v. "Colonel Baron Takeichi Nishi," as retrieved in May 2020 from the following url: https://www.facebook.com/notes/world-war-ii-history/colonel-baron-takeichi-nishi/10150114201296034/.

16. *Iwo Jima,* by Newcomb, p. 323.

17. Ibid., "Place Names on Iwo" appendix, p. 323.

18. *CosCom Learn Japanese*, s.v. "Japanese Numbers," as retrieved in June 2020 from the following url: https://www.coscom.co.jp/learnjapanese101/wordcategory/basicwords_numbers.html.

19. *Jisho Japanese Dictionary,* s.v. "futatsu," as retrieved in June 2020 from the following url: https://jisho.org/search/futatsu.

20. Report of Intelligence: Iwo Jima Operation," ACofS G-2, FMFPac, maps.

21. Bartley, *Iwo Jima: Amphibious Epic,* picture caption, p. 180.

22. Ibid., pp. 121, 122.

23. Ibid., p. 179.

24. "Report of Intelligence: Iwo Jima Operation," by ACofS G-2, FMFPac, p. 37; "Western Pacific Operations," by Garand and Strowbridge, vol. IV, part VI, chapt. 11, p. 695; Bartley, *Iwo Jima: Amphibious Epic,* p. 206

25. Bartley, *Iwo Jima: Amphibious Epic,* p. 29.

26. "Report of Intelligence: Iwo Jima Operation," by ACofS G-2, FMFPac, maps.

Chapter 13: *Best Way to Breach a Modern Defense "Line"*

1. "Western Pacific Operations," by Garand and Strowbridge, vol. IV, part VI, chapt. 4, p. 516; "Iwo Jima: The Most Horrific of All World War II Battles," from Warfare Hist. Network; "Remember the Sacrifice of Iwo Jima Veterans," *Marines TV;* "Report of Intelligence: Iwo Jima Operation," by ACofS G-2, FMFPac. p. 30; Alexander, *Closing In,* p. 2.

2. *Strategic Rifleman,* by Poole, chapt. 13.

3. William S. Lind, in Foreword to *One More Bridge to Cross,* by H. John Poole (Emerald Isle, NC: Posterity Press, 1998).

4. "The 4-Man SAS Patrol," *Elite UK Forces* on line, n.d.; Michael Lee Lanning and Dan Cragg, *Inside the VC and the NVA: The Real Story of North Vietnam's Armed Forces* (New York: Ivy Books, 1992), pp. 211, 212.

5. *Super-Squad,* by Poole, epilogue.

6. "Battle Cry" (Hollywood: Warner Bros., February 1955), videocassette, 149 minutes.

7. Company C, 1st Battalion, 23rd Marines website, "Approach and Capture of Iwo Jima"; Alexander, *Closing In,* p. 23.

8. *Gung Ho,* by Poole, chapt. 10.

9. *Home of the Heroes,* s.v "Daniel J. Marini.'

10. *Wikipedia Encyclopedia*, s.v. "Joseph R. Julian."

11. Bartley, *Iwo Jima: Amphibious Epic,* picture on p. 133.

12. *Home of the Heroes,* s.v. "Willard F. Gabriel."

13. *Wikipedia Encyclopedia*, s.v. "Jack Lummus."

14. Ibid., s.v. "Wilson D. Watson" and "Douglas T. Jacobson."

15. *Home of the Heroes,* s.v. "Luther C. Crabtree."

16. Ibid., s.v. "Harold C. Boehm."

17. Alexander, *Closing In,* p. 37.

18. *Home of the Heroes*, s.v. "Nolen M. Garrett."

19. Ibid., s.v. "Claude E. Lauderdale."

20. Bartley, *Iwo Jima: Amphibious Epic,* p. 37.

21. *Home of the Heroes,* s.v. "Louis Komnenich."

22. Ibid., s.v. "Cleo S. Danford, Jr."

23. Ibid., s.v. "Salvador Vargas."

24. Alexander, *Closing In,* p. 50.

25. Memorandum for the record by H.J. Poole, after serving as a rifle company commander in Vietnam for eight months.

26. "Western Pacific Operations," by Garand and Strowbridge, vol. IV, part VI, chapt. 1, p. 455.

27. Ross, *Iwo Jima: Legacy of Valor,* pp. 293, 294.

28. Ibid., p. 294.

29. *Warfighting,* MCDP 1, pp. 42, 43; Xiaobing Li, *China's Battle for Korea: The 1951 Spring Offensive* (Bloomington, IN: Indiana Univ. Press, 2014), pp. xxv-xxvi.

30. *AZ Quotes,* s.v. "George S. Patton," as retrieved in October 2018 from the following url: https://www.azquotes.com/quote/226589.

31. *Gung Ho*, by Poole, chapt. 10.

32. *Okinawa: The Last Battle,* by Roy E. Appleman, James M. Burns, Russell A. Gugeler, and John Stevens, *United States Army in World War II Series* (Washington, D.C.: U.S. Army's Ctr. of Mil. Hist., 2000), pp. 322, 323.

33. *Gung Ho,* by Poole, chapt. 9; *Marines in the Central Solomons,* by Rentz, chapt. 4, "From Rice to Bairoko," pp. 113, 114.

34. *Gung Ho,* by Poole, chapt. 10.

35. Maj.Gen. Robert H. Scales, U.S. Army (Ret.), "Infantry and National Priorities," *Armed Forces Journal,* December 2007, pp. 14-17, 45.

36. "North Korea May Have Aided *Hezbollah,* LTTE - U.S. Report," by Arshad Mohammed, *Reuters* on line, 13 December 2007; "North Koreans Assisted *Hezbollah* with Tunnel Construction," Jamestown Foundation, *Terrorism Focus,* vol. III, issue 30, August 2006; *ABC's Nightly News,* 28 July 2006.

37. Nicholas Blanford, Daniel McGrory, and Stephen Farrell, "Tactics That Have Kept the Middle East's Most Powerful Army at Bay," *The Times* (U.K.), 10 August 2006.

38. "Why Did Armored Corps Fail in Lebanon," by Hanan Greenberg, *Israeli News,* 30 August 2006.

39. Edward Cody and Molly Moore, "Analysts Attribute *Hezbollah's* Resilience to Zeal, Secrecy and Iranian Funding," *Washington Post,* 14 August 2006.

40. "North Korea May Have Aided *Hezbollah,* LTTE - U.S. Report," by Mohammed; Blanford, McGrory, and Farrell, "Tactics That Have Kept the Middle East's Most Powerful Army at Bay"; Anthony H. Cordesman, "Iran's Revolutionary Guards, the Al Quds Force, and Other Intelligence and Paramilitary Forces," working draft (Washington, D.C.: Ctr. for Strategic and Internat. Studies, 16 August 2007), p. 10.

41. Blanford, McGrory, and Farrell, "Tactics That Have Kept the Middle East's Most Powerful Army at Bay."

42. Nicholas Blanford, "Hizbullah's Resilience Built on Years of Homework," *Christian Science Monitor,* 11 August 2006, p. 4.

43. *Global Warrior: Averting WWIII,* by H. John Poole (Emerald Isle, NC: Posterity Press, 2011), chapt. 3; "Why Did Armored Corps Fail in Lebanon," by Greenberg.

44. *Gung Ho,* by Poole, p. 138-141; *Dan Marsh's Marine Raider Page,* created by former member of 4th Raider Battalion, s.v. "Okinawa: The Last Battle."

45. "Finnish Tactics—Small Units," a U.S. Intelligence Report on Finnish Tactics in WWII, *Tactical and Technical Trends,* no. 6, 27 August 1942.

46. *Handbook on Japanese Forces,* TM-E 30-480, p. 117.

Chapter 14: *If Iwo Were Stormed Today*

1. "Book of Daniel," *New King James Bible* (N.p.: Christian Art Publishers; 2015), chapt. 3, verse 25.
2. Alexander, *Closing In,* p. 34; quote from Erskine at the start of part four.
3. Gy.Sgt. Rodney Walker USMC (Ret.) [career infantryman and former super-squad member], in conversations with author around 2018; Memorandum for the record by H.J. Poole, after a 28-year career in the USMC.
4. *Marine Rifle Company / Platoon* FMFM 6-4 (Quantico, VA: Marine Corps Develop. and Educ. Cmd., 1978), p. 5.
5. *The Speadhead: WWII History of the 5th Marine Division,* by Connor, p. 43.
6. "Encyclopedic Dictionary," *The New American Bible* (N.p.: Catholic World Press/World Bible Publishers, 1987).
7. Alexander, *Closing In,* p. 43.
8. *Wikipedia Encyclopedia,* s.v. "flamethrower"; "Prohibitions or Restrictions on the Use of Incendiary Weapons (Protocol III)," U.S. Dept. of State, agreement signed at Geneva in 1980.
9. "Drone-Syria.mp4," from "Drones Pinpoint and Take Out ISIS Fighters . . . ," by Gabriel Zamfir and Rachel Bishop, *Mirror* on line (UK), 29 August 2017.
10. Wright, *Iwo Jima 1945: The Marines Raise the Flag on Mount Suribachi,* p. 22, in *Wikipedia Encyclopedia,* s.v. "Battle of Iwo Jima."
11. Marine infantryman, quoted in Ross, *Iwo Jima: Legacy of Valor,* p. 135.
12. Merritt E. Benson, "Seventeen Days at Iwo Jima," *Marine Corps Gazette,* February 2000, p. 32.
13. Ross, *Iwo Jima: Legacy of Valor,* p. 262.
14. 2ndLt. Clifton A. Cascaden, "Selfless Devotion Immortalized by Iwo Jima Memorial," *Camp Lejeune Globe* (North Carolina), 8 October 1999, p. 8A.
15. *So Sad to Fall in Battle,* by Kakehashi, pp. 67, 68.
16. *Super-Squad,* by Poole, chapt. 14.
17. "Reflections on Iwo Jima: The Japanese Tunnels," interview of Col. Dave Severance USMC (Ret.).
18. "Ground Penetrating Radar FAQ," USRadarInc., n.d.

19. "Night Fever: NATO's 1999 Structured Technology Demonstration Showcases Night-Vision Systems," *Armed Forces Journal International,* August 1999, p. 37.

20. "Geology and Ground-Water Resources of Iwo Jima," by Swenson; *Iwo Jima,* by Newcomb, p. 4; Alexander, *Closing In,* map on page 5.

21. "Into the Meatgrinder," chapt. 6, of *Battle for Iwo Jima,* by Leckie.

22. Ibid.

23. *Arrest and Trial,* ABC TV, 17 November 2000.

24. Col. William D. Siuru, Jr., USAF (Ret.), "Emerging Technologies," *Marine Corps Gazette,* January 1999, p. 41.

25. "Why Doesn't the US Army Develop Thermobaric Weapons," *Forbes Magazine* on line, Quora submission, 21 March 2017.

26. Ross, *Iwo Jima*, p. 163.

27. *So Sad to Fall in Battle,* by Kakehashi, p. 68.

28. "U.S. Drops Largest Non-Nuclear Bomb on Islamic State," by Justin Sink, *Bloomberg News* on line, 13 April 2017; "About the Giant Bomb the U.S. Used in Afghanistan," by Anne Cronin, *Bloomberg News* on line, 13 April 2017.

29. Ibid.

30. "N. Korea Military Tactics in a War with U.S.: A Strategy of Massive Retaliations Against U.S. Attacks," by Han Ho Suk (Seoul: Ctr. for Korean Affairs [www.onekorea.org], 24 April 2003), English abstract by Jeff Rense.

31. *North Korea's Hidden Assets,* by H. John Poole (Emerald Isle, NC: Posterity Press, 2018), chapt. 4.

32. Ibid.

33. *Wikipedia Encyclopedia,* s.v. "M67 Flame Thrower Tank."

34. "Prohibitions or Restrictions on the Use of Incendiary Weapons (Protocol III)," U.S. Dept. of State; *Wikipedia Encyclopedia,* s.v. "Mark 77 bomb."

35. "Target: Iwo Jima – Sidebar: June '00 American History," by Christine Techky, *American History Magazine,* n.d.; "Western Pacific Operations," by Garand and Strowbridge, vol. IV, part VI, pp. 644, 711; Bartley, *Iwo Jima: Amphibious Epic,* pp. 9, 148; "The 147th Infantry Regiment at Iwo Jima," by Nash, p. 14; *Investigating Iwo,* ed. Robertson, p. 37; "Fierce Fighting Footage from the Battle of Iwo Jima (1945)," War Archives, British Pathe, Youtube video presentation, 30 August 2011.

36. "Editorial: The Terrible Cost of Victory on Iwo Jima," *Richmond Times Dispatch,* 19 February 2020.

37. *Super-Squad,* by Poole, chapts. 8 and 14.

38. "Windtalkers" (Hollywood: MGM, 2002), DVD, 134 min.

Glossary

A.k.a.	Also known as	An alias name for someone or something
Amtrac	Amphibious Tractor	Tracked type of U.S. amphibious landing vehicle, same as LVT.
BLT	Battalion Landing Team	Battalion-sized amphibious assault unit.
CO	Commanding Officer	Senior unit leader.
CR	Collection Request	Map elevation.
D-2	Military staff billet	Divisional intelligence officer.
D-Day	D in D-Day stands for Day	Start of important invasion or operation
DoD	Department of Defense	Headquarters for all U.S. military services
DMZ	Demilitarized Zone	No-mans land between warring nations.
DUKW	D means 1942, U means utility (amphibious), K means all-wheel drive, W means two powered rear axles.	Type of military wheeled amphibious landing-craft.
E-1	U.S. military pay grade	That for a Marine Private.

E-2	U.S. military pay grade	That for a Marine PFC.
E-7	U.S. military pay grade	That for a Marine Gunnery Sergeant.
FAE	Fuel Air Explosive	Thermobaric weapon
FMFPac	Fleet Marine Forces Pacific	Parent command of Marine divisions on Iwo Jima
G-2	Military staff billet	Intelligence officer
GC	Grid Coordinate	Numeric scale for map making.
GI	Government Issue	Colloquial term for low-ranking member of U.S. Armed Forces
GPR	Ground Penetration Radar	Electromagnetic waves that detect anomalies just below the earth's surface.
HQMC	Headquarters Marine Corps	Top echelon of USMC.
IJA	Imperial Japanese Army	Ground forces of Japan.
ISIS	Islamic State in Iraq and Syria	Middle East terrorist organization.
LCM	Landing Craft Mechanized	U.S. amphibious landing craft.
LCU	Landing Craft Utility	U.S. amphibious landing craft
LCVP	Landing Craft, Vehicle, Personnel	Higgins Boat.

LP	Listening Post	Nighttime OP.
LST	Landing Ship Tank	U.S. amphibious landing craft
LT	Landing Team	Amphibious assault unit.
LVT	Landing Vehicle Tracked	Same as Amtrac.
MLR	Main Line of Resistance	Linear formation of defensive positions.
MOAB	Massive Ordnance Air Blast	FAE bomb to produce massive overpressure and oxygen depletion
MoH	Medal of Honor	Highest U.S. award for valor.
MOP	Massive Ordnance Penetrator	Bunker-buster bomb.
NCO	Noncommissioned Officer	Junior enlisted military leader.
NVA	North Vietnamese Army	Hanoi-directed ground forces.
OP	Observation Post	Forward sentry site.
PFC	Private First Class	Second-echelon enlisted rank.
POW	Prisoner of War	Captured combatant.
RCT	Regimental Combat Team	Reinforced infantry regiment.
SAS	Special Air Service	British commandos.
2GW	2nd Generation Warfare	Focus is on killing enemy soldiers; same as traditional U.S. style (attrition warfare).

SNCO	Staff Noncommissioned Officer	Senior enlisted military leader.
SP	Sentry Post	Two-man outpost.
TA	Target Area	Map code
TF	Task Force	Mission-oriented grouping of units.
3GW	3rd Generation Warfare	Bypasses foe's man-power concentrations to more easily get at his command and logistics networks.
TNT	Trinitrotoluene	Explosive used in WWII satchel charge.
UDT	Underwater Demolition Team	U.S. frogmen.
U.N.	United Nations	Worldwide alliance of countries.
U.S.	United States	America.
USMC	U.S. Marine Corps	America's amphibious force in readiness.
UW	Unconventional Warfare	Attacking through infiltration and defending through ambushes; part of 3GW.
VAC	V Amphibious Corps	U.S. Navy and Marine TF
VC	Viet Cong	Enemy guerrillas in Vietnam
WWI	World War I	First global conflict.
WWII	World War II	Second global conflict

Bibliography

U.S. Government Publications, News Releases, and Videos

Alexander, Col. Joseph H. *Closing In: Marines in the Seizure of Iwo Jima.* Marines in World War II Commemorative Series. Washington, D.C.: History and Museums Division, Headquarters Marine Corps, 1994. As retrieved in January 2020 from the following urls: https://ia902707.us.archive.org/20/items/ClosingIn/ClosingIn.pdf; https://www.nps.gov/parkhistory/online_books/npswapa/extcontent/usmc/pcn-190-003131-00/pcn-190-003131-00/index.htm.

The Amphibians Came to Conquer: The Story of Admiral Richmond Kelly Turner. By George Carroll Dyer. Department of the Navy. United States Marine Corps. FMFRP 12-109-II. Volume 2. As retrieved in January 2020 from the following url: https://books.google.com/books?id=6ga4abz2UgMC&pg=PA1023&lpg=PA1023&dq=9000+marines+landed+in+25+minutes+at+iwo+jima&source=bl&ots=XX4VS9BDVs&sig=ACfU3U0PCx9kzxMeJm-qVUi7qt134PnutQ&hl=en&ppis=_c&sa=X&ved=2ahUKEwiCv_K98pTnAhVuUN8KHTF0AUMQ6AEwEHoECAwQAQ#v=onepage&q=9000%20marines%20landed%20in%2025%20minutes%20at%20iwo%20jima&f=false.

Bartley, Lt.Col. Whitman S. *Iwo Jima: Amphibious Epic.* Washington, D.C.: Historical Branch, Headquarters Marine Corps, 1954. As retrieved in December 2019 from the following urls: https://archive.org/stream/IwoJimaAmphibiousEpic/IwoJimaAmphibiousEpic_djvu.txt; and https://books.google.com/books?id=cDRNAQAAMAAJ&pg=PA114&lpg=PA114&dq=search+of+kangoku+rock+at+iwo+jima&source=bl&ots=ezHO7g4v2h&sig=ACfU3U3qtlQud5eiVecIN1na6RmIwuaITg&hl=en&ppis=_c&sa=X&ved=2ahUKEwjw0M6h89XmAhXSKM0KHf62DA4Q6AEwCnoECA0QAQ#v=onepage&q=kangoku&f=false.

"The Battle of Iwo Jima—Marines Remember Their Heritage." Interview of Col. Dave Severance USMC (Ret.), CO of E/2/25. *Marines TV,* YouTube video, 4 March 2011. As retrieved in January 2020 from the following url: https://www.youtube.com/watch?v=epVX8zlU3-4.

Cascaden, 2ndLt. Clifton A. "Selfless Devotion Immortalized by Iwo Jima Memorial." *Camp Lejeune Globe* (North Carolina), 8 October 1999.

From Makin to Bougainville: Marine Raiders in the Pacific War. By Maj. Jon T. Hoffman, USMCR. Marines in World War II Commemorative Series. Washington, D.C.: History and Museums Division, Headquarters Marine Corps, n.d. As retrieved in May 2020 from the following url: https://www.usmcu.edu/Portals/218/From%20Makin%20to%20Bougainville-Marine%20Raiders%20in%20the%20Pacific%20War%20PCN%2019000313000.pdf?ver=2018-10-30-102328-017.

"Geology and Ground-Water Resources of Iwo Jima." By Frank A. Swenson. GSA Bulletin (1948) 59 (10): 995–1008, 1 October 1948. As retrieved in January 2020 from the following url: https://pubs.geoscienceworld.org/gsa/gsabulletin/article-abstract/59/10/995/4188/GEOLOGY-AND-GROUND-WATER-RESOURCES-OF-IWO-JIMA?redirectedFrom=fulltext.

Handbook on Japanese Military Forces. TM-E 30-480. Washington, D.C.: U.S. War Department, 1944. Reprint, Baton Rouge, LA: LSU Press, 1991.

"Into the Meatgrinder." Chapter 6. *Battle for Iwo Jima.* By Robert Leckie. N.p.: I Books, 2004. As retrieved in February 2020 from the following url: https://erenow.net/ww/battle-for-iwo-jima/6.php.

Investigating Iwo: The Flag Raisings in Myth, Memory, & Esprit de Corps. Contributing editor, Breanne Robertson. Quantico, VA: Marine Corps History Division, 2019. As retrieved in January 2020 from the following url: https://www.usmcu.edu/Portals/218/Investigating%20Iwo_WEB2.pdf.

"Iwo Jima: The Unnecessary Battle." By Maj. Jonathan T. Baker. Master's Thesis. Quantico, VA: Command and Staff College, Marine Corps University, 25 March 2011. As retrieved in January 2020 from the following url: https://apps.dtic.mil/dtic/tr/fulltext/u2/a600230.pdf.

Kennedy, President John F. Quote from his 1962 West Point graduation speech. In "Unconventional Warfare Fundamentals," at *The Irregular Warrior* website.

Marine Rifle Company / Platoon. FMFM 6-4. Quantico, VA: Marine Corps Development and Education Command, 1978.

Marines in the Central Solomons. By Maj. John N. Rentz. "United States Marine Corps Historical Monograph." Washington, D.C.: Historical Branch, Headquarters Marine Corps, 1952. Chapter 4, "From Rice to Bairoko."

"Marine Who Led Famous Charge up Mountain on Iwo Jima Remembered." By Matthew Schehl. *Marine Corps Times,* 17 February 2016. As retrieved in March 2020 from the following url: https://www.marinecorpstimes.com/news/your-marine-corps/2016/02/17/marine-who-led-wwii-charge-up-mount-suribachi-on-iwo-jima-dies/.

"Memories of Iwo Jima. Interview of four Iwo Jima veterans. *Marines TV,* 26 March 2018. As retrieved in January 2020 from the following url: https://www.youtube.com/watch?v=qUrePwJmsCc.

"The 147th Infantry Regiment at Iwo Jima." By Douglas E. Nash, Sr. *Army History Magazine,* Fall 2017. U.S. Army Center of Military History. As retrieved in January 2020 from the following url: https://history.army.mil/armyhistory/AH105(W).pdf.

"Prohibitions or Restrictions on the Use of Incendiary Weapons (Protocol III)." U.S. Department of State. Agreement signed at Geneva in 1980. As retrieved in February 2017 from the following url: http://www.state.gov/documents/organization/190579.pdf.

"Reflections on Iwo Jima: The Japanese Tunnels." Interview of Col. Dave Severance USMC (Ret.), CO of E/2/25. *Marines TV,* Youtube video, 26 March 2011. As retrieved in December 2019 from the following url: https://www.youtube.com/watch?v=NVkrhvqlyOY.

"Remember the Sacrifice of Iwo Jima Veterans." *Marines TV*. Youtube video, 14 March 2012. As retrieved in January 2020 from the following url: https://www.youtube.com/watch?v=UON0HSadM4E.

"Report of Intelligence: Iwo Jima Operation." Headquarters, Expeditionary Troops, Task Force 56. ACofS G-2, Fleet Marine Forces Pacific memorandum to the Commanding General of 1 April 1945. As retrieved in March 2020 from the following url: http:// www.ibiblio.org/hyperwar/PTO/Iwo/ComTaskForFiftySixIwoJima/TaskForFiftySix-EnclosureC.pdf.

"31st [Naval Construction Battalion] Spearhead." N.p., n.d. As retrieved in February 2020 from the following url: 31 NCB 1942-45.pdf.

"To the Shores of Iwo Jima," U.S. Department of the Navy, 22 February 2017. With U.S. National Archives footage from Department of the Navy, Office of the Chief of Naval Operations, Naval Observatory, 1942-9/18/1947, of the movie of that name in 1945. Series of Navy Motion Picture Film Productions, ca. 1939 - ca. 1947. Record Group 80 of General Records of the Department of the Navy, 1804 - 1983. National Archives Identifier: 13089. Local Identifier: 80-MN-5124. National Archives Catalog: https://catalog.archives.gov/id/13089. As retrieved in January 2020 from the following url: https://www.youtube.com/watch?v=h4E9t2D1obo. Clearest version of modified film at https://www.youtube.com/watch?v=qG3HNHBFBzU. Original unedited film may be at https://www.youtube.com/watch?v=4er8u9vd-Pw.

The United States Marines on Iwo Jima: The Battle and the Flag Raisings. By Bernard C. Nalty and Danny J. Crawford. Washington, D.C.: History and Museums Division, Headquarters Marine Corps, 1995. As retrieved in May 2020 from the following url: *https://www.marines.mil/Portals/1/Publications/The%20United%20*States%20Marines%20On%20Iwo%20Jima_The%20Battle%20and%20the%20Flag%20Raisings%20%20PCN%2019000316600.pdf.

Warfighting. MCDP 1. Washington, D.C.: Headquarters Marine Corps, 1997.

"Western Pacific Operations," *History of Marine Corps Operations in World War II.* By George W. Garand and Truman R. Strowbridge. Historical Branch, G-3 Division, Headquarters Marine Corps, 1971. Volume IV. Part VI, chapters 1 and 4, as retrieved in January 2020 from the following urls: https://www.ibiblio.org/hyperwar/USMC/IV/USMC-IV-VI-1.html and https://www.ibiblio.org/hyperwar/USMC/IV/USMC-IV-VI-4.html. And Chapter 11 as retrieved from this url: https://www. ibiblio.org/hyperwar/USMC/IV/USMC-IV-VI-11.html. Whole book available from this url: https://www.ibiblio.org/hyperwar/USMC/IV/index.html.

Civilian Publications

Analytical Studies and Websites

AZ Quotes. From its website, https://www.azquotes.com.

Battle Story: Iwo Jima 1945. By Andrew Rawson. Cheltenham, U.K.: The History Press, July 2012. As retrieved in December 2019 from the following urls: (1) https://erenow.net/ww/battle-story-iwo-jima-1945/5.php; and (2) https://books.google.com/books?id=VdcGCgAAQBAJ&pg=PA50&lpg=PA50&dq=The+Marines+had+to+work+their+way+forward+under+covering+fire+until+they+were+close+enough+to+push+the+charge+through+the+embrasure.&source=bl&ots=YUVYzN8j8h&sig=ACfU3U0WGgk46cnAprfGxX-WExnhAYp_0w&hl=en&ppis=_c&sa=X&ved=2ahUKEwjO4P2UuNPmAhWVbc0KHYYmBlYQ6AEwAHoECAgQAQ#v=onepage&q=The%20Marines%20had%20to%20work%20their%20way%20forward%20under%20covering%20fire%20until%20they%20were%20close%20enough%20to%20push%20the%20charge%20through%20the%20embrasure.&f=false.

Company C, 1st Battalion, 23rd Marines website. "Approach and Capture of Iwo Jima." As retrieved through this url: http://www.c123rd.com/our-wwii-history/chapter-7-approach-and-capture-of-iwo-jima.

Cordesman, Anthony H. "Iran's Revolutionary Guards, the Al Quds Force, and Other Intelligence and Paramilitary Forces." Working draft. Washington, D.C.: Center for Strategic and International Studies, 16 August 2007.

CosCom Learn Japanese. From its website, https://www.coscom.co.jp/.

Dan Marsh's Marine Raider Page. Created by former member of 4th Raider Battalion. As retrieved in November 2011 from its website, www.usmcraiders.com.

Dower, John W. *War Without Mercy: Race and Power in the Pacific War*. New York, Pantheon, 1987. As retrieved in April 2020 from the following url: https://books.google.com/books?id=8himI4wNnxEC&q=Senjinkun#v=snippet&q=Senjinkun&f=false.

Ennis. By Laurie J. Wilson and Peggy Holland Rankin, N.p.: Arcadia Publishing, February 2009. As retrieved in January 2020 from the following url: https://books.google.com/books?id=iBalPTCYYNgC&pg=PA111&lpg=PA111&dq=9000+marines+at+iwo+jima&source=bl&ots=MAXhwUf0bZ&sig=ACfU3U1JQd8kArjNDHH0czc_9QaIdXrolA&hl=en&ppis=_c&sa=X&ved=2ahUKEwiZ4YCS7pTnAhUvmuAKHfOlAg04FBDoATAKegQIChAB#v=onepage&q=9000%20marines%20at%20iwo%20jima&f=false.

The First Battalion of the 28th Marines on Iwo Jima: A Day-by-Day History from Personal Accounts and Official Reports, with Complete Muster Rolls. By Robert E. Allen. London: McFarland & Company, 1999. As retrieved in June 2020 from the following url: https://books.google.com/books?id=8BOBCgAAQBAJ&printsec=frontcover#v=onepage&q&f=false.

From Uraga to San Francisco: A Century of Japanese Diplomacy, 1853-1952. By Hisahiko Okazaki. Japan: Echo, 2007. As retrieved in March 2020 from the following url: https://books.google.com/books/about/From_Uraga_to_San_Francisco_A_Century_of.html?id=se0-AQAAIAAJ.

Give Me Fifty Marines Not Afraid to Die. By John K. Wells. N.p.: Ka-Well Enterprises, 1995.

Global Warrior: Averting WWIII. By H. John Poole. Emerald Isle, NC: Posterity Press, 2011.

Guardian Joe: How Less Force Helps the Warrior. By H. John Poole. Emerald Isle, NC: Posterity Press, 2018.

Gung Ho: The Corps' Most Progressive Tradition. By H. John Poole. Emerald Isle, NC: Posterity Press, 2012.

Hall of Valor Project. From its website, https://valor.militarytimes.com.

Home of the Heroes. From its website, https://homeofheroes.com.

I'm Staying with My Boys: The Heroic Life of Sgt. John Basilone, USMC. By Jim Prosser with Jerry Cutter [Basilone nephew]. New York: St. Martin's Griffin, 2004.

Iwo Jima: The Account of the Epic Battle That Turned the Tide of World War II. By Richard F. Newcomb. N.p: Griffin, May 2002. As retrieved in March 2020 from the following url: https://books.google.com/books?id=iPvmNzkuLn4C&pg=PA323&lpg=PA323&dq=#v=onepage&q&f=false.

"Japanese Defenses and Fortifications Tawara, Iwo, Jima, and Okinawa 1943-1945." By J. David Rogers and Karl F. Hasserman Chair. University of Missouri at Rolla, n.d. As retrieved in February 2020 from the following url: https://web.mst.edu/~rogersda/umrcourses/ge342/Japanese%20Island%20Defenses%201943-45.pdf.

Jisho Japanese Dictionary. At its website, https://jisho.org.

Lanning, Micheel Lee and Dan Cragg. *Inside the VC and the NVA: The Real Story of North Vietnam's Armed Forces*. New York: Ivy Books, 1992.

The New American Bible. N.p.: Catholic World Press/World Bible Publishers, 1987.

New King James Bible. N.p.: Christian Art Publishers, 2015.

"Operation Detachment: The Battle for Iwo Jima February - March 1945." *History of War,* n.d. As retrieved in March 2020 from the following url: http://www.historyofwar.org/articles/battles_iwojima.html.

North Korea's Hidden Assets. By H. John Poole. Emerald Isle, NC: Posterity Press, 2018.

Ross, Bill D. *Iwo Jima: Legacy of Valor*. New York: Vintage, 1986.

The Second World War. Edited by Jeremy Black. Volume III, "The Japanese War 1941-1945." As retrieved in March 2020 from the following url: https://books.google.com/books/about/The_Second_World_War.html?id=FAokDwAAQBAJ.

Shaw, John and the editors of Time-Life Books. *Red Army Resurgent* Chicago: Time-Life Books, 1979.

So Sad to Fall in Battle: An Account of War Based on General Tadamichi [Kuribayashi]. By De Kumiko Kakehashi. Novato, CA: Presidio Press, 2009. Reprint edition. As retrieved in January 2020 from these urls: https://books.google.com/books?id=GTUVD9qN7dQC&printsec=copyright#v=onepage&q&f=false; https://books.google.fr/books?id=GTUVD9qN7dQC&pg=PA62&lpg=PA62&dq=senda+sadasue&source=bl&ots=OnYSYs3ks8&sig=ACfU3U3vqV6hX0OghpDrlMm0YkkEYO4Ibg&hl=fr&sa=X&ved=2ahUKEwjtjtzMuZLlAhUB6OAKHYy_A-MQ6AEwEXoECAkQAQ#v=onepage&q=senda%20sadasue&f=false.

The Spearhead (5th Marine Division's unofficial website). From the following url: http://www.5thmarinedivision.com.

The Spearhead: The World War II History of the 5th Marine Division. By Howard M. Connor. Washington, D.C.: Infantry Journal Press, 1950.

Strategic Rifleman: Key to More Moral Warfare. By H. John Poole. Emerald Isle, NC: Posterity Press, 2014.

Super-Squad: The Now Missing Component. By H. John Poole. Emerald Isle, NC: Posterity Press, January 2020.

The Third Marine Division. By Robert A. Aurthur and Kenneth Cohlmia. Edited by Robert T. Vance. Washington, D.C.: Infantry Journal Press, 1948. As retrieved in January 2020 from the following url: https://babel.hathitrust.org/cgi/ssd?id=uc1.31822016006256;page=ssd;view=plaintext;seq=267;num=255.

The Tiger's Way: A U.S. Private's Best Chance at Survival. By H. John Poole. Emerald Isle, NC: Posterity Press, 2003.

Warner, Philip. *Japanese Army of World War II*. Men-at-Arms Series. Color plates by Michael Youens. London: Osprey Publications Limited, 1972.

Weatherbase. From its website, https://www.weatherbase.com.

World War II Medal of Honor Recipients. London: Osprey, n.d. As retrieved in June 2020 from the following url: https://books.google.com/books?id=HtkbDAAAQBAJ&pg=PA51&lpg=PA51&dq=was+tony+stein+a+former+raider&source=bl&ots=ZPK09cD5iE&sig=ACfU3U081SXtqprkV67C80AwlYZNwmfMVw&hl=en&sa=X&ved=2ahUKEwibt9j1uYbqAhUZTDABHVAQCUIQ6AEwAXoECAwQAQ#v=onepage&q=was%20tony%20stein%20a%20former%20raider&f=false

Xiaobing Li. *China's Battle for Korea: The 1951 Spring Offensive*. Bloomington, IN: Indiana University Press, 2014.

Letters, E-Mail, and Verbal Conversations

Gray, Jerry (retired U.S. Navy Chief Petty Officer). In conversation with author around January 2020.

Miller, Maj.Gen. Edward J. USMC (with replacement draft on Iwo Jima as a new lieutenant). In conversation with author around 1975.

Walker, Gy.Sgt. Rodney USMC (Ret.) [career infantryman and former super-squad member]. In multiple conversations with author over the years.

Videotapes, Movies, DVDs, TV Programs, Slide Shows, CDs, Illustrations

Arrest and Trial. ABC TV, 17 November 2000.

"Battle Cry." Hollywood: Warner Brothers, February 1955. Videocassette, 149 minutes.

"Drone-Syria.mp4." From "Drones pinpoint and take out ISIS fighters . . . ," by Gabriel Zamfir and Rachel Bishop. *Mirror* on line (UK), 29 August 2017. As retrieved in February 2020 from the following url: https://www.mirror.co.uk/news/world-news/drones-used-kill-isis-fighters-11074008.

"Fierce Fighting Footage (1945)." War Archives. British Pathe, Youtube video presentation, 30 August 2011. As retrieved in May 2020 from the following url: https://www.youtube.com/watch?v=8rfsRNeiing.

"Jack Lucas and the Battle of Iwo Jima." *The History Guy*. Youtube video presentation, 26 February 2020. As retrieved in February 2020 from the following url: https://www.youtube.com/watch?v=JKM0OiQZe4g.

"Japanese Codetalkers" *In Search of History*. History Channel, 30 March 1999.

Sands of Iwo Jima. Hollywood: Republic Pictures, 1988. Videocassette.

"Situation Map of Iwo Jima." As retrieved in February 2020 from the following url: https://bostonraremaps.com/inventory/situation-map-iwo-jima-1944/.

"Vehicle path and combat-shattered vegetation in the rugged terrain of northern Iwo Jima," WWII Database. Photograph, 21 April 1945. As retrieved in January 2020 from the following url: https://ww2db.com/image.php?image_id=k3155.

What Cameramen Saw Through Their Lenses, 85:00. A Tribute to WWII Combat Cameramen of Japan Series. Tokyo: Nippon TV, 3 August 1995.

"Windtalkers," Hollywood: MGM, 2002. DVD, 134 minutes.

Newspaper, Magazine, and Website Articles

"About the Giant Bomb the U.S. Used in Afghanistan." By Anne Cronin. *Bloomberg News* on line, 13 April 2017.

"Absolutely Brutal (and Smart) Tactics in Native American Warfare." By Justin Andress. *Ranker,* 4 February 2019. As retrieved in May 2020 from the following url: https://www.ranker.com/list/native-american-tactics/justin-andress.

Assessing the Conventional Balance: The 3:1 Rule and Its Critics." By John J. Mearsheimer. *International Security.* Volume 13, number 4, Spring 1989. As retrieved in April 2020 from the following url: https://www.jstor.org/stable/2538780?seq=1.

"Battle of Iwo Jima." By Joseph Alexander. *History Net,* n.d. As retrieved in January 2020 from the following url: https://www.historynet.com/battle-of-iwo-jima.htm.

"Battle of Iwo Jima 19 Feb 1945 - 26 Mar 1945." By C. Peter Chen. *World War II Database,* n.d. As retrieved in February 2020 from the following url: https://ww2db.com/battle_spec.php?battle_id=12.

Benson, Merritt E. "Seventeen Days at Iwo Jima." *Marine Corps Gazette,* February 2000.

"The Black Sands of Iwo Jima." American Veteran's Center, n.d. As retrieved in January 2020 from the following url: https://www.americanveteranscenter.org/avc-media/magazine/avq/avq-issue-ii/the-black-sands-of-iwo-jima/.

Blanford, Nicholas. "Hizbullah's Resilience Built on Years of Homework." *Christian Science Monitor,* 11 August 2006.

Blanford, Nicholas, and Daniel McGrory, and Stephen Farrell. "Tactics That Have Kept the Middle East's Most Powerful Army at Bay." *The Times* (U.K.), 10 August 2006.

Cody, Edward and Molly Moore. "Analysts Attribute *Hezbollah's* Resilience to Zeal, Secrecy and Iranian Funding." *Washington Post,* 14 August 2006.

"Editorial: The Terrible Cost of Victory on Iwo Jima," *Richmond Times Dispatch,* 19 February 2020. As retrieved in February 2020 from the following url: https://www.richmond.com/opinion/editorial/editorial-the-terrible-cost-of-victory-on-iwo-jima/article_d1fca82d-771b-5fc2-bfe3-8ef91d915d44.html.

"Excellence in Land Navigation: The 1stLt. Keith Wells Award." Marine Corps Reserve Association website. As retrieved in March 2020 from the following url: "https://usmcra.org/page-1819736.

"Finnish Tactics—Small Units." A U.S. Intelligence Report on Finnish Tactics in WWII. *Tactical and Technical Trends.* Number 6, 27 August 1942.

"The 4-Man SAS Patrol." *Elite UK Forces,* n.d. As retrieved in March 2020 from the following url: https://www.eliteukforces.info/special-air-service/4-man-patrol/.

"Ground Penetrating Radar FAQ." USRadarInc., n.d. As retrieved in August 2017 from the following url: http://www.usradar.com/about-ground-penetrating-radar-gpr/faq/.

"Gyokusai or 'Shattering like a Jewel': Reflection on the Pacific War." By Hiroaki Sato. *Asia-Pacific Journal,* 1 February 2008. As retrieved in April 2020 from the following url: https://apjjf.org/-Hiroaki-Sato/2662/article.html.

"IJN Submarine I-44: Tabular Record of Movement" By Bob Hackett & Sander Kingsepp, n.d. As retrieved in March 2020 from the following url: http://www.combinedfleet.com/I-44.htm.

"Iwo Jima and Okinawa: Death at Japan's Doorstep." *National WWII Museum* website, 10 July 2017. As retrieved in January 2020 from the following url: https://www.nationalww2museum.org/war/articles/iwo-jima-and-okinawa-death-japans-doorstep?gclid=CjwKCAiAx_DwBRAfEiwA3vwZYmDouev1LCMDcm5_DiGMOgEWpPqhbx6zrF7D2XYGVw1R8SMWvrA8ZRoCTc4QAvD_BwE.

"Iwo Jima Journal; A Pacific Isle That Can't Quite Rest in Peace." By Sheryl Wudunn. *New York Times,* 15 March 1995. As retrieved in January 2020 from the following url: https://www.nytimes.com/1995/03/15/world/iwo-jima-journal-a-pacific-isle-that-can-t-quite-rest-in-peace.html.

"Iwo Jima: The Most Horrific of All World War II Battles." By Warfare History Network. *The National Interest,* 15 December 2018. As retrieved in January 2020 from the following url: https://nationalinterest.org/blog/buzz/iwo-jima-most-horrific-all-world-war-ii-battles-38897?page=0%2C1.

"Japanese Organisation in WWII." As retrieved in November 2018 from this url: http://www-solar.mcs.st-and.ac.uk/~aaron/WW2ORG/jap.html.

"The Legend of Suicide Charley: Company C, 1st Battalion, 7th Marines, and the Battle for Henderson Field, 24–25 October 1942." By Major Gary Cozzens, USMCR (Ret). *Marine Corps History,* Summer 2017. As retrieved in May 2020 from the following url: https://www.usmcu.edu/Portals/218/MCH%20Vol%203%20No%201.pdf?ver=2018-10-16-115407-260.

Lind, William S. In Foreword of *One More Bridge to Cross,* by H. John Poole. Emerald Isle, NC: Posterity Press, 1998.

"Marine Who Led Famous Charge up Mountain on Iwo Jima Remembered." *USMC Life,* 18 February 2016. As retrieved in March 2020 from the following url: https://www.usmclife.com/2016/02/marine-who-led-famous-charge-up-mountain-on-iwo-jima-remembered/.

"Night Fever: NATO's 1999 Structured Technology Demonstration Showcases Night-Vision Systems." *Armed Forces Journal International,* August 1999.

"N. Korea Military Tactics in a War with U.S.: A Strategy of Massive Retaliations Against U.S. Attacks." By Han Ho Suk. Seoul: Center for Korean Affairs (www.onekorea.org), 24 April 2003. English abstract by Jeff Rense. As retrieved in May 2018 from this url: http://www.rense.com/general37/nkorr.htm. Original translation may have come from this url: http://information clearinghouse.literati.org/article3099.htm.

"North Korea May Have Aided *Hezbollah,* LTTE - U.S. Report." By Arshad Mohammed. *Reuters* on line, 13 December 2007. As retrieved in April 2018 from the following url: https://in.reuters.com/article/idINIndia-30964520071213.

"North Koreans Assisted *Hezbollah* with Tunnel Construction." Jamestown Foundation. *Terrorism Focus.* Volume III, issue 30, August 2006.

"Obituaries: Colonel Charles Waterhouse USMCR." *The Star-Ledger* (Newark, NJ), 18 November 2013. As retrieved in December 2019, from the following url: https://obits.nj.com/obituaries/starledger/obituary.aspx?pid=168081934.

"Return to Iwo Jima: 75 years After US-Japan Battle, What Is the Island Like Today?" By Julian Ryall. This Week in Asia. *South China Morning Post,* 2 June 2019. As retrieved in December 2019 from the following url: https://www.scmp.com/week-asia/geopolitics/article/3012683/return-iwo-jima-75-years-after-us-japan-battle-what-island.

"Russo-Japanese War: Japan's First Big Surprise." By Robert M. Citino, *History Net,* n.d. As retrieved in April 2020 from the following url: https://www.historynet.com/russo-japanese-war-japans-first-big-surprise.htm.

Scales, Maj.Gen. Robert H., U.S. Army (Ret.). "Infantry and National Priorities." *Armed Forces Journal,* December 2007.

"A Season In Hell: Iwo Jima 50th Anniversary." By Nathan Miller. *Baltimore Sun,* 19 February 1995. As retrieved in February 2020 from the following url: https://www.baltimoresun.com/news/bs-xpm-1995-02-19-1995050098-story.html.

"The Sinking of the Japanese Submarine I-1 off of Guadalcanal and the Recovery of Its Secret Documents." By Netisha. N.p., 31 October 2017. Posted in *Military Records, Navy, Marines, & Coast Guard, World War II,* by Dr. Greg Bradsher, Senior Archivist at the National Archives at College Park. As retrieved in March 2020 from the following url: https://text-message.blogs.archives.gov/2017/10/31/the-sinking-of-the-japanese-submarine-i-1-off-of-guadalcanal-and-the-recovery-of-its-secret-documents/.

Siuru, Col. William D., Jr., USAF (Ret.). "Emerging Technologies." *Marine Corps Gazette,* January 1999.

"Sulfur Island." By P.J. O'Rourke. *The Atlantic,* June 2004. As retrieved in February 2020 from the following url: https://www.theatlantic.com/magazine/archive/2004/06/sulfur-island/305332/.

"Sun Tzu Quotes." *Goodreads,* n.d. As retrieved in April 2020 from the following url: https://www.goodreads.com/author/quotes/1771.

"Target: Iwo Jima – Sidebar: June '00 American History." By Christine Techky. *American History Magazine,* n.d. As retrieved in April 2020 from the following url: https://www.historynet.com/target-iwo-jima-sidebar-june-00-american-history.htm.

"U.S. Drops Largest Non-Nuclear Bomb on Islamic State." By Justin Sink. *Bloomberg News* on line, 13 April 2017.

"Why Did Armored Corps Fail in Lebanon." By Hanan Greenberg. *Israeli News,* 30 August 2006. As retrieved in 2017 from this url: http://www.ynetnews.com/articles/0,7340,L-3297431,00.html.

"Why Doesn't the US Army Develop Thermobaric Weapons?" *Forbes Magazine* on line. Quora submission, 21 March 2017. As retrieved in January 2020 from the following url: https://www.forbes.com/sites/quora/2017/03/21/why-does-the-us-army-not-develop-thermobaric-missiles/#698812973fdf.

"The WW II Japanese Army on Iwo Jima had 1000 Korean laborers building the caves, tunnels and fortifications." By Grant Lee. *Quora,* 2 June 2017. As retrieved in June 2020 from the following url: https://www.quora.com/The-WW-II-Japanese-Army-on-Iwo-Jima-had-1000-Korean-laborers-building-the-caves-tunnels-and-fortifications-What-became-of-these-Koreans.

About the Author

After 28 years of commissioned and noncommissioned infantry service, John Poole retired from the United States Marine Corps in April 1993. While on active duty, he studied small-unit tactics for nine years: (1) six months at the Basic School in Quantico (1966); (2) seven months as a rifle platoon commander in Vietnam (1966-67); (3) three months as a rifle company commander at Camp Pendleton (1967); (4) five months as a regimental headquarters company (and camp) commander in Vietnam (1968); (5) eight months as a rifle company commander in Vietnam (1968-69); (6) five and a half years as an instructor with the Advanced Infantry Training Company (AITC) at Camp Lejeune (1986-92); and (7) one year as the Staff Noncommissioned Officer in Charge (SNCOIC) of the 3rd Marine Division Combat Squad Leaders Course (CSLC) on Okinawa (1992-93).

While at AITC, he developed, taught, and refined courses on maneuver warfare, land navigation, fire support coordination, call for fire, adjust fire, close air support, M203 grenade launcher, movement to contact, daylight attack, night attack, infiltration, defense, offensive Military Operations in Urban Terrain (MOUT), defensive MOUT, Nuclear/Biological/Chemical (NBC) defense, and leadership. While at CSLC, he further refined the same periods of instruction and developed others on patrolling.

He has completed all of the correspondence school requirements for the Marine Corps Command and Staff College, Naval War College (1,000-hour curriculum), and Marine Corps Warfighting Skills Program. He is a graduate of the Camp Lejeune Instructional Management Course, the 2nd Marine Division Skill Leaders in Advanced Marksmanship (SLAM) Course, and the East-Coast School of Infantry Platoon Sergeants' Course.

In the 25 years since retirement, Poole has researched the small-unit tactics of other nations and written 18 other books: (1) *The Last Hundred Yards,* a squad combat study based on the consensus opinions of 1,200 formal school NCOs and casualty statistics of AITC and CSLC field trials; (2) *One More Bridge to Cross,* a treatise on enemy proficiency at short range and how to match it; (3) *Phantom Soldier,* an in-depth look at the highly deceptive Asian style of war; (4) *The Tiger's Way,* the fighting styles of Eastern fireteams and soldiers; (5) *Tactics of the Crescent Moon,* insurgent procedures in Palestine, Chechnya, Afghanistan, and Iraq; (6) *Militant Tricks,* how Muslim insurgents fight in an urban environment; (7) *Terrorist Trail,* tracing the jihadists in Iraq back to their home countries; (8) *Dragon Days,* an unconventional warfare technique manual; (9) *Tequila Junction,* how to fight narco-guerrillas; (10) *Homeland Siege,* countering the 4GW assault on America by a foreign power's organized-crime proxies; (11) *Expeditionary Eagles,* how to outmaneuver the Taliban; (12) *Global Warrior,* forestalling WWIII through tiny-contingent activity; (13) *Gung Ho,* how supporting arms are not needed to take strongpoint matrices; (14) *Strategic Rifleman,* how to produce tiny semi-autonomous combat elements; (15) *Afrique,* with Africa as a model how U.S. intelligence is being gathered incorrectly for certain adversaries; (16) *Sinoland,* how the U.S. has been under 4GW attack from China since 9/11; (17) *Guardian Joe,* how—through more skilled infantry squads—the Pentagon can use less force to more easily win a modern war; and (18) *North Korea's Hidden Assets,* 13 little-known wartime capabilities of what has been erroneously called the Hermit Kingdom.

Since retirement, Poole has traveled extensively in both Communist and Islamist worlds to study the infantry tactics of America's enemies. He has done research in Mainland China (twice), North Korea, Vietnam, Cambodia, Pakistan (twice), Bangladesh, Malaysia, Iran, Lebanon, Turkey, the Emirates, Egypt, Sudan, Tanzania, Zambia, South Africa, Morocco, and Venezuela. Over the course of his lifetime, he has visited scores of other nations on all five continents.

As of April 2014, John Poole had conducted multi-day training sessions (on advanced squad tactics) at 41 (mostly Marine) battalions, nine Marine schools, and seven special-operations units from all four U.S. service branches.

Between early tours in the Marine Corps (from 1969 to 1971), he served as a criminal investigator with the Illinois Bureau of Investigation (IBI). After attending the State Police Academy for several months in Springfield, he was assigned to the IBI's Chicago office. There, he worked mostly on general criminal and drug cases.

Several periods in Poole's life have qualified him to accurately mirror the viewpoints of America's true fighters— its junior enlisted infantrymen. In 1965, his first official orders were for "Private Poole." From 1992 to 1993, he served as SNCOIC of a divisional squad leaders school. Then, almost all of the 1100 students who attended his "post-retirement" training sessions were E-5's and below.

Poole is one of the very few Americans with an extensive enough background to know how to fix the ongoing tactical shortfall within the U.S. military. After some 60 multi-day training sessions at different active-duty units, he has developed, tested, and refined its solution (through a "bottom-up" training supplement). He has also become adept at the now virtually lost art of field training light infantrymen (those who are able do well on their own in the absence of orders).

Name Index

F

G

H

I

J

K

L

T

U

No entries.

V

W

X

No entries.

Y

No entries.

Z

No entries.